Apartheid's Leviathan

NEW AFRICAN HISTORIES

SERIES EDITORS: JEAN ALLMAN, ALLEN ISAACMAN, DEREK R. PETERSON, AND CARINA RAY

David William Cohen and E. S. Atieno Odhiambo, *The Risks of Knowledge*

Belinda Bozzoli, *Theatres of Struggle and the End of Apartheid*

Gary Kynoch, *We Are Fighting the World*

Stephanie Newell, *The Forger's Tale*

Jacob A. Tropp, *Natures of Colonial Change*

Jan Bender Shetler, *Imagining Serengeti*

Cheikh Anta Babou, *Fighting the Greater Jihad*

Marc Epprecht, *Heterosexual Africa?*

Marissa J. Moorman, *Intonations*

Karen E. Flint, *Healing Traditions*

Derek R. Peterson and Giacomo Macola, editors, *Recasting the Past*

Moses E. Ochonu, *Colonial Meltdown*

Emily S. Burrill, Richard L. Roberts, and Elizabeth Thornberry, editors, *Domestic Violence and the Law in Colonial and Postcolonial Africa*

Daniel R. Magaziner, *The Law and the Prophets*

Emily Lynn Osborn, *Our New Husbands Are Here*

Robert Trent Vinson, *The Americans Are Coming!*

James R. Brennan, *Taifa*

Benjamin N. Lawrance and Richard L. Roberts, editors, *Trafficking in Slavery's Wake*

David M. Gordon, *Invisible Agents*

Allen F. Isaacman and Barbara S. Isaacman, *Dams, Displacement, and the Delusion of Development*

Stephanie Newell, *The Power to Name*

Gibril R. Cole, *The Krio of West Africa*

Matthew M. Heaton, *Black Skin, White Coats*

Meredith Terretta, *Nation of Outlaws, State of Violence*

Paolo Israel, *In Step with the Times*

Michelle R. Moyd, *Violent Intermediaries*

Abosede A. George, *Making Modern Girls*

Alicia C. Decker, *In Idi Amin's Shadow*

Rachel Jean-Baptiste, *Conjugal Rights*

Shobana Shankar, *Who Shall Enter Paradise?*

Emily S. Burrill, *States of Marriage*

Todd Cleveland, *Diamonds in the Rough*

Carina E. Ray, *Crossing the Color Line*

Sarah Van Beurden, *Authentically African*

Giacomo Macola, *The Gun in Central Africa*

Lynn Schler, *Nation on Board*

Julie MacArthur, *Cartography and the Political Imagination*

Abou B. Bamba, *African Miracle, African Mirage*

Daniel Magaziner, *The Art of Life in South Africa*

Paul Ocobock, *An Uncertain Age*

Keren Weitzberg, *We Do Not Have Borders*

Nuno Domingos, *Football and Colonialism*

Jeffrey S. Ahlman, *Living with Nkrumahism*

Bianca Murillo, *Market Encounters*

Laura Fair, *Reel Pleasures*

Thomas F. McDow, *Buying Time*

Jon Soske, *Internal Frontiers*

Elizabeth W. Giorgis, *Modernist Art in Ethiopia*

Matthew V. Bender, *Water Brings No Harm*

David Morton, *Age of Concrete*

Marissa J. Moorman, *Powerful Frequencies*

Ndubueze L. Mbah, *Emergent Masculinities*

Judith A. Byfield, *The Great Upheaval*

Patricia Hayes and Gary Minkley, editors, *Ambivalent*

Mari K. Webel, *The Politics of Disease Control*

Kara Moskowitz, *Seeing Like a Citizen*

Jacob Dlamini, *Safari Nation*

Alice Wiemers, *Village Work*

Cheikh Anta Babou, *The Muridiyya on the Move*

Laura Ann Twagira, *Embodied Engineering*

Marissa Mika, *Africanizing Oncology*

Holly Hanson, *To Speak and Be Heard*

Paul S. Landau, *Spear*

Saheed Aderinto, *Animality and Colonial Subjecthood in Africa*

Katherine Bruce-Lockhart, *Carceral Afterlives*

Natasha Erlank, *Convening Black Intimacy in Early Twentieth-Century South Africa*

Morgan J. Robinson, *A Language for the World*

Faeeza Ballim, *Apartheid's Leviathan*

Apartheid's Leviathan

Electricity and the Power of Technological Ambivalence

Faeeza Ballim

OHIO UNIVERSITY PRESS

ATHENS, OHIO

Ohio University Press, Athens, Ohio 45701
ohioswallow.com
© 2023 by Ohio University Press
All rights reserved

To obtain permission to quote, reprint, or otherwise reproduce or distribute material from Ohio University Press publications, please contact our rights and permissions department at (740) 593-1154 or (740) 593-4536 (fax).

Printed in the United States of America
Ohio University Press books are printed on acid-free paper ∞ ™

Library of Congress Cataloging-in-Publication Data available upon request
Names: Ballim, Faeeza, author.
Title: Apartheid's leviathan : electricity and the power of technological ambivalence / Faeeza Ballim.
Other titles: New African histories series.
Description: Athens : Ohio University Press, 2023. | Series: New African histories | Includes bibliographical references and index.
Identifiers: LCCN 2022051214 (print) | LCCN 2022051215 (ebook) | ISBN 9780821425183 (paperback) | ISBN 9780821425176 (hardcover) | ISBN 9780821447963 (pdf)
Subjects: LCSH: Electric power production—Economic aspects—South Africa. | Electric power production—Political aspects—South Africa. | Apartheid—Economic aspects—South Africa. | South Africa—Economic conditions.
Classification: LCC HD9685.S62 B46 2023 (print) | LCC HD9685.S62 (ebook) | DDC 333.793/20968—dc23/eng/20230130
LC record available at https://lccn.loc.gov/2022051214
LC ebook record available at https://lccn.loc.gov/2022051215

For my parents, Yunus Ballim and Naseera Ali

Contents

	Acknowledgments	ix
	Introduction	1
Chapter 1	The Unlikely Exploitation of the Waterberg	26
Chapter 2	The Taming of the Waterberg	43
Chapter 3	Eskom and the Turning of the Tide	59
Chapter 4	Contested Neoliberalism	73
Chapter 5	Labor and Belonging in Lephalale	89
Chapter 6	The Medupi Power Station	106
	Conclusion	123
	Notes	131
	Bibliography	153
	Index	161

Acknowledgments

As with any piece of writing that is long in the making, I owe a huge debt of gratitude to countless people who have shown me good humor and generosity throughout.

My greatest debt is to Keith Breckenridge, who supervised the dissertation out of which this book arose and who uniquely portended the important role of Eskom and Medupi in postapartheid South Africa as well as the implications for postcolonial societies more generally. Colleagues at the Wits Institute for Social and Economic Research (WiSER) created a thriving intellectual space through regular seminars and conversations in the corridor and provided the institutional support for the PhD. Max Bolt, Belinda Bozzoli, Catherine Burns, Adilla Deshmukh, Najibha Deshmukh, Sarah Emily-Duff, Pamila Gupta, Shireen Hassim, Jonathan Klaaren, Charne Lavery, Achille Mbembe, Hlonipha Mokoena, Sarah Nuttall, and Antina von Schnitzler each offered great encouragement and intellectual support. In addition, fellow PhD students at WiSER offered camaraderie, including Robyn Bloch, Candice Jansen, Christi Kruger, Emery Kalema, Ruth Sacks, Ellison Tijera, Rene van der Wiel, and Natasha Vally.

I am additionally grateful to members of the various conferences and workshop that arose out of the Mellon-funded University of Michigan and WiSER collaboration. These gatherings, held in Ann Arbor, Durban, and Maputo, offered scholarly engagement and personable conversations about my work. Members include Kevin Donovan, Paul Edwards, Gabrielle Hecht, Iginio Gagliardone, Euclides Gonzales, Emma Park, Derek Peterson, and Lynne Thomas. Members of the Law, Organization, Science and Technology research group in 2021, convened by Richard Rottenburg and including Lorenz Gosch, David Kananizadeh, Laura Matt, Georges Eyenga, Johannes Machinya, Bronwyn Kotzen, and Jessica Breakey, read and offered valuable commentary on parts of this book. I am deeply indebted to Richard

Rottenburg in particular for generously sharing insights into the voluminous science and technology studies scholarship as a fellow coeditor of the Translating Technologies book project. I must also give thanks to Allen Isaacman for his patient help and guidance for a green author such as myself.

Members of the Wits history department and the History Workshop have been hugely important in my career and research, including Prinisha Badassy, Peter Delius, Noor Nieftagodien, Arianna Lissoni, Franziska Rueedi, Antonette Gouws, Clive Glaser, Andrew Macdonald, Maria Suriano, and Sekiba Lekgoathi.

My colleagues at the University of Johannesburg have been hugely supportive of the writing of this book and a great source of friendship: Greg Barton, Brett Bennett, Natasha Erlank, Nafisa Essop-Sheik, Gerald Groenewald, Juan Klee, Khumisho Moguerane, Stephen Sparks, Sithembile Thusi, and Thembisa Waetjen.

I must also acknowledge the support and encouragement of my friends Nurina Ally, Fatima Vally, Dasantha Pillay, Sameera Munshi, Naadirah Munshi, Michelle Hay, Anne Heffernan, Andrew Bowman, and Judy and Tom Heffernan. Thanks to Shireen, Sattar, Nazreen, and Amina for their unflappable hospitality during countless trips to Limpopo. And to my family, Imraan and Raiza, and my parents, Yunus and Naseera, for the endless support and good humor during the many years it took for this book to come together.

Apartheid's Leviathan

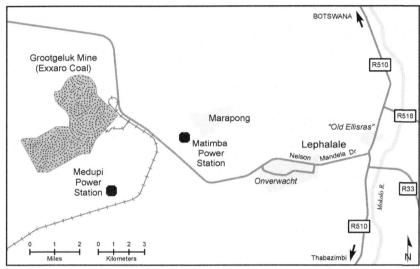

Map of Lephalale town and surrounding area. Created by Brian Balsley.

Introduction

AT THE beginning of 2018, South Africa's national electricity provider, the Electricity Supply Commission of South Africa (Eskom) threatened to sink the fortunes of the South African government. Eskom was indebted to the tune of R450 billion (35 billion in USD), and most of this debt was backed by government guarantees. Eskom was then in the midst of a liquidity crunch, and in the event that it failed to meet its repayment obligations, the government would have to assume responsibility, a cost that the fiscus could scarce afford.[1] Eskom managed to regain its liquidity in subsequent months and lumbered on, still remaining a corporation of concern. The following year, at an event hosted by the investment firm Goldman Sachs, South African president Cyril Ramaphosa told a gathering of foreign investors that Eskom would not be privatized. Eskom was "just too big to fail," Ramaphosa said. "It holds the fortunes at an economic level and social level of our country in its hands."[2]

In a country where a celebrated culture of popular protest hastened the demise of the apartheid regime and continues to challenge the practices of the postapartheid government, Eskom—an essentially technological corporation—is an unexpected threat to the fortunes of the South African government. The intimacy of its relationship with the South African government was etched into the principles of its founding under the Electricity Act of 1922. Prime Minister Jan Smuts had encouraged the formation of Eskom, and then in 1928, amid general government enthusiasm for state corporations, the Pact government, under General J. B. M. Hertzog, oversaw the creation of the steel manufacturing corporation Iscor. The two state corporations were organizationally and juridically distinct from each other and from the South African government, and because they were essentially technological corporations, engineers dominated their upper echelons. Eskom in particular developed a momentum of its own as the twentieth century wore on. It grew from strength to strength so that, according to an economist based at the University of Witwatersrand, Stuart Jones, the replacement value of its assets was an estimated R60,000 million ($3,200 million)[3] in August 1986, an amount that was larger than the market value of the country's gold mines, which stood at R51,000 million ($2,800 million).[4] In other words, Eskom was richer in investment than South Africa's most important activity of the twentieth century, one that had supplied the bulk of the tax revenue used to build the capacity of the state bureaucracy and for which electricity had first been generated.

Over time, Eskom surpassed any auxiliary role it might have initially played as subordinate to the electricity requirements of the gold mines, and it acted in close cooperation with the successive governments of the twentieth century. These governments were consumed with the project of white supremacy in various degrees of intensity before 1994, and, under apartheid, realized the most ambitious program of racial segregation in the world. While the state corporations were organizationally autonomous from the government, their activities appeared to follow the imperatives of the latter in an uncanny fashion.

This book is, firstly, concerned with understanding this seemingly contradictory relationship between the different South African governments of the twentieth century and the state corporations. The story told here begins in the 1960s, setting the scene for a period of heightened repression in South Africa that fits the mold of James Scott's authoritarian high modernism. But rather than viewing the technological state corporations as tools in the arsenal of authoritarian rulers, this book reveals their ambivalent relationship—one that can be characterized as both autonomous and immersive. Insights from the field of science and technology studies are particularly valuable in attempting to understand this relationship. Scholars working in this field have complicated the idea of intentionality, dwelling instead on the interstitial space between command and action. Scholars have also remarked on the promiscuity of technologies, whether in purpose or scope, and the variation in the user experience of already existing infrastructure.[5] This book shows that two of the largest technical systems in the world—South Africa's national electricity provider Eskom and its national steel manufacturer Iscor—created infrastructures under the protection of an authoritarian government but generated contradictory politics internally and in the societies in which they operated. This ambivalence allowed Eskom to survive, with a singular tenacity, into the era of political and economic liberalization, since it could be repurposed to serve the ends of the democratic government.[6] In the late 2000s, Eskom attempted to resuscitate its construction network, but with limited success. Now, far from being a docile instrument of government command, it has drawn the government to the brink of bankruptcy.

Secondly, Eskom and Iscor are conceived here as writhing leviathans, made up of disparate elements that are both human and nonhuman, and with the ability to move through time and space in a coordinated fashion. Such a conceptualization complicates our existing understanding of the major political and economic transformation of the African continent during the second half of the twentieth century: a transformation that went from authoritarian governments committed to state-led development to neoliberalism and democracy. The oil crisis of 1973 is generally considered the beginning of the end of the state-led development project across the African continent and the start of the rise of the governmental austerity, which entailed the diminution in the size and capacity of the state bureaucracy, characteristic of the neoliberal era.[7] This book demonstrates that rather than destroying the South African state corporations, the oil crisis initiated a relative austerity with a distinct tenacity on the part of the state corporation engineers. This peculiar combination kept the engine of the developmental project running and led, ultimately, to the exploitation of the Waterberg coalfields. In the 1970s these coalfields were the last remaining coal frontier in South Africa. Despite the fact that Eskom, as a large-scale, monopolistic state corporation, did not comfortably align with neoliberal principles of competition, it escaped privatization in the 1980s and 1990s. And in 2007, Eskom was drawn into renewed government exuberance in spending on infrastructure projects when it began construction on the Medupi power station, a cornerstone of Eskom's resurgent commitment to expanding its electricity generation capacity.

Lastly, this book demonstrates that South Africa's transition from authoritarian rule to democracy meant for Eskom, in part, a transition from dogged certainty to one of interminable uncertainty. The engineers' tenacity in the face of the austerity of the 1970s rose out of their adherence to their long-term predictions of demand and supply. This adherence was, in turn, necessary to ensure the survival of White South Africa, and the defensive,

Introduction ⁓ 3

nation-building imperative endowed a certain cohesion to the relationship between the apartheid government and the state corporations. Democracy institutionalized contestation and offered up a splintered imaginary of the future. In the 1980s, Eskom's sure-footedness was already threatened by seemingly uncontrollable elements, such as trade unions and environmental activists. Eskom adapted to the changing imperatives of neoliberalism and of racial transformation in the early 1990s. And the Medupi power station, begun in 2007, became a construction project capacious enough to absorb the different imperatives of the democratic era.

AUTHORITARIAN HIGH MODERNISM

The story told here begins in the 1960s, a decade in which the apartheid government fully embraced the power of scientific planning to realize racial segregation. The first prime minister of the apartheid regime, D. F. Malan, was wary of the state corporations because British imperial sympathizer Jan Smuts was instrumental in their creation. But Hendrik Verwoerd, prime minister of South Africa from 1958 to 1966, more fully embraced scientific principles to realize racial segregation within his vision of "grand apartheid," and he enjoyed a close relationship with the leadership of the state corporations. The Sharpeville Massacre of 21 March 1960, which was widely reported in the foreign press, marked the start of international hostility to the apartheid regime and raised the very real threat that international sanctions would be imposed against South Africa. The government saw scientific planning as essential to the project of national industrialization. Industrial development and the growth of the manufacturing sector would ensure that South Africa reduced its dependence on imported goods in the likely event that sanctions were imposed.

The 1960s in South Africa thus marks the beginning of an extreme period of the authoritarian high modernism that James Scott describes in his seminal work on the brutal, state-driven interventions that occurred around the world in the nineteenth and twentieth centuries.[8] In Scott's formulation, authoritarian governments brutally intervened in societies within their ambit of control, ignoring the opinions of the local populace in favor of a modernization that rested on the authority of scientific expertise. In South Africa, this governmental praxis is clearly seen in the activities of the Group Areas Board (GAB), the official body responsible for racial segregation, which carried out forced removals of people in racially mixed neighborhoods that were labeled slums. Forced removals occurred in earnest during the second half of the twentieth century, affecting both commercial and residential

4 ～ *Apartheid's Leviathan*

districts in large cities and in the tiniest of towns across South Africa.[9] The GAB instructed the people forcibly removed from their homes to reside in "townships" that it had developed as racially delimited mini-towns in close proximity to a major urban center that then became the preserve of Whites. These townships usually consisted of a small business district, a residential area, and bare fields that served as public parks and sports grounds. More generally, these instances of state-driven social engineering valorized scientific experts because of their ability to achieve social and economic modernization. In Scott's formulation, however, scientists and engineers appear as docile handmaidens of governments. The organizational and professional loyalties of the technical experts and planners under discussion are vague, and the state, scientists, and engineers appear as undifferentiated conspirators subsumed within the overarching vehicle of authoritarian high modernism.

Eskom and Iscor can be considered agents of authoritarian high modernism since Scott's analysis is not restricted to scientists and engineers within the government bureaucracy. Scott writes that at times the task of authoritarian high modernism fell to agencies with "quasi-governmental powers" and the ability to enact large-scale technological interventions.[10] Eskom has certainly been portrayed in a similar vein in the South African historiography, as having acted in concert with governments to further their authoritarian aims. Renfrew Christie, for example, argues that Eskom served only the interests of racial capitalism, which in turn strengthened the fiscus of successive colonial and apartheid governments.[11] Electricity from Eskom powered the machines that enabled the gold mines to reach ever-deeper levels and cast off their dependence on skilled African labor. This, in turn, created the conditions for the maintenance of the "color bar," whereby skilled White operatives managed a cheap, low-skilled African labor force. Similarly, Ben Fine and Zavareh Rustomjee argue that the minerals-energy complex, which derives its profits from the export of minerals, has been the dominant driver of the South African economy in the twentieth century.[12] For Fine and Rustomjee, Eskom served the interests of the mines—chiefly the gold mines—creating a "minerals-energy complex" that has prevented the prosperity of economic activity that occurred outside of it. This remains the most influential interpretation of twentieth-century South African economic (and institutional) history.

But Eskom (and Iscor) did not exist solely to serve the needs of the gold mines. Together with South African Railways and Harbours, they formed the foundation of industrialization by providing cheap electricity

Introduction ⁓ 5

and steel to both the mines and to the infant manufacturing sector. And while driven by the demands of the mines, both contained a developmental purpose from their origin, which was focused on improving the lot of the poor White population.[13] In the years preceding the creation of Eskom and Iscor, Jan Smuts had experienced the full might of the White working class during a series of insurrectionary strikes that he temporarily subdued in 1914. This created a "crisis of legitimacy"[14] for the government, and Smuts resolved to play a more interventionist role in industrial development to create employment for Whites. In this way, the state corporations contained a distinctive socioeconomic mission that was tied to the government's efforts to protect White workers, especially after 1922. Their developmental role would endure in various political forms for the rest of the twentieth century.

This dual role of Eskom and Iscor continued into the 1960s and fused with the apartheid government's defensive effort to promote industrialization. But industrialization also provided the basis for the beginning of worker solidarity and the bitterly contested development of African trade unions—especially the Metal and Allied Workers Union and its peers—from the 1970s onward. Worker protest and organization, in turn, played an important role in the protest culture of the late apartheid period and the eventual dismantling of the apartheid state.[15] This ambivalent role of the state corporations—one that supplied the steel and electricity for the apartheid regime while simultaneously planting the seeds for its challenge—is further explored in this book.

As the South African government embarked on the project of grand apartheid in the 1960s, elsewhere on the continent, newly independent countries rapidly descended into authoritarian rule, casting off the trappings of democracy that departing colonial powers had hastily introduced.[16] Postcolonial leaders took up the mantle of the developmental project that colonial governments had implemented toward the end of their rule to quell African unrest. Both colonial and postcolonial leaders hoped that electricity generated from new hydroelectric dams would simultaneously modernize mineral exploitation and kick-start national industrialization. For postcolonial leaders, electricity lay at the heart of their promise to deliver the fruits of modernization to the majority of the citizenry. Some of these dams were constructed during the colonial period, such as the Kariba Dam of the Central African Federation.[17] Others were built in countries still under colonial rule later than the general pattern across the continent, such as the Cahora Bassa Dam of Mozambique, completed in 1975, which entailed the forced

6 ⁓ *Apartheid's Leviathan*

removals of indigenous peoples from the banks of the Lower Zambesi River before construction began.[18]

These large dams were expensive to build and a huge financial burden on the governments that commissioned them. For example, Mobutu Sese Seko, president of Zaire (present-day Democratic Republic of the Congo) from 1965 to 1997, hoped that the Inga-Shaba hydroelectric project in the Congo would provide the electricity to intensify copper mining in the copper-rich Shaba region. The resultant cost overruns and fatal delays of the Shaba Dam were major contributors to the financial collapse of the postcolonial Congolese state in the 1980s.[19] More successful examples of hydroelectric projects on the continent include President Kwame Nkrumah's Akosombo Dam in Ghana, which Nkrumah hoped would power a giant aluminum plant that processed naturally occurring bauxite. In each case, as in South Africa, electricity formed the basis of economic and social modernization, while frequently rooted in the need to service the mines for the export of minerals.

SCIENCE, TECHNOLOGY, AND AFRICAN STUDIES

Studies of African politics and the African state have neglected a concerted engagement with the technological facet of power. While highlighting the importance of technology and infrastructure, these studies have not taken into account the analytical implications of the vitality of technology. Notable among these is Jeffrey Herbst's study of the way that states project their authority from their location at the capital to the peripheral regions of the territory within the borders of the nation-state.[20] Herbst argues that this process in Africa differed from the centripetal forces at work in European nation-states, which had engaged in warfare with their neighbors for centuries. In African countries, roads were essential to overcoming geographical constraints and reaching far-flung communities, but the paucity of roads and other communication infrastructure deterred the effective transmission of state authority. The challenge that incumbent leaders face in extending state power throughout the national territory is a recurring theme in studies of African politics.

In his study of the 2009 construction of the Merowe Dam in Sudan under the presidency of Omar al-Bashir, Harry Verhoeven notes the essential "civilizing" mission that al-Bashir envisioned developmental projects performing, uniting a country historically riven by violent cleavages.[21] Al-Bashir hoped that the dam would strengthen the links between the capital and peripheral regions of the territory within the national borders, thus bolstering

Introduction ⁓ 7

the political fortunes of the military-Islamist al-Ingaz regime. Despite these efforts, South Sudan seceded in 2011, frustrating the government's efforts to maintain territorial integrity. Iginio Gagliardone has documented the role that information and control technologies have played in Ethiopia in transmitting political orders from the country's capital to the bureaucrats stationed in the various administrative districts. This transmission of governmental authority mirrored the regimental style of the ruling party, the Ethiopian People's Ruling Democratic Front, which had previously borne arms as the country's liberation party.[22]

Historians of South Africa have recently focused attention on the role of scientific institutions in the country's history, challenging the idea of a trajectory of political development driven by human action.[23] Keith Breckenridge has examined the importance of technological failure in Prime Minister Verwoerd's decision-making during his term in office. In an effort to resolve one of the most important challenges of racial segregation—controlling African mobility in the nominally White urban areas—Verwoerd ordered the creation of a national fingerprint database called the *bewysburo*.[24] The failure of this scheme, Breckenridge suggests, gave rise to government efforts to make the African homelands durable as the last remaining option for total segregation. In a recent book, *The Scientific Imagination in South Africa,* William Beinart and Saul Dubow have highlighted the autonomous role of scientific developments and institutions, separate from the governments of their time, as well as their role in maintaining group identities of belonging and exclusion.[25] While recognizing the dual, ambivalent role of science and technology, their work presents various instances of scientific ingenuity in South African history, and they do not concertedly examine the relationship between the South African governments and the scientists and engineers.

A small but significant body of scholarship has examined this relationship through the lens of "technopolitics." The term is best understood in its scholarly context, rather than as a generally applicable definition, and the books of three authors who thoroughly engage it are discussed here: Gabrielle Hecht, Timothy Mitchell, and Antina von Schnitzler. Historian of technology Thomas Hughes, writing in 1986, characterized the relationship between technology and society as a "seamless web," one in which humans and technology are mutually constituted and ultimately inseparable.[26] In studying nuclear development in France after the Second World War, Gabrielle Hecht expanded the bounds of the technological system to include the French nation-building project. Driven by state-linked corporations,

8 ～ *Apartheid's Leviathan*

the Électricité de France and the Commissariat al'Energie Atomique, the development of nuclear power assisted with the reconstruction of national identity in a country devastated by war and shorn of its former imperial glory. By rhetorically separating the sphere of technology from the sphere of politics, engineers at the state-linked corporations cloaked their activities in an apolitical guise. At the same time, the particular properties of the technology delimited the options available to politicians and so, Hecht argues, "technology cannot be reduced to politics."[27]

In *Rule of Experts: Egypt, Techno-Politics, Modernity,* a book focused on colonial and postcolonial Egypt, Timothy Mitchell describes technopolitics as an "alloy" that is composed of both the human and the nonhuman, but organized so that human intention appears to overpower and drive the nonhuman.[28] Mitchell offers a model for gauging change over time by demonstrating the ultimately messy enactment of colonial experts' preconceived plans and their continued reformulation in the face of unpredictable, particularly material, elements.

Lastly, Antina von Schnitzler's *Democracy's Infrastructure: Techno-Politics and Protest after Apartheid* details the adoption of a small and seemingly mundane object in South Africa, the prepaid meter. Johannesburg city engineers installed prepaid meters in African townships in the 1980s to alleviate the crisis of nonpayment. The nonpayment was in turn a product of resident distrust of municipal authorities and recurrent unrest in the townships. Prepaid meters continue to be used in the postapartheid period and are a rallying object for protests against the lack of service delivery. Von Schnitzler argues that, rather than a "conduit for power," they constitute a "political terrain," one in which popular protest, the rights of the citizenry, and governmental intervention all play out.[29]

An important characteristic of a technopolitical intervention, as in the studies discussed above, is its open-endedness. This goes against the grain of positivist political theory, which presumes that the presence of certain conditions will give rise to predictable outcomes through the operation of discernible laws. By virtue of its technical complexity, technology holds the potential to entirely defy the will of the human operator and so scuttle the purpose of the latter, whether for better or worse. It is thus difficult to identify pure zones of political and technological action. Government bureaucracies are inextricably bound up with technologies and infrastructures that enable communications, record keeping, and security, to name a few. Politics is conceived here as composed of both human and nonhuman agents that are not restricted to the governmental sphere or to the machinations

Introduction ⁓ 9

of political parties. Rather, politics is composed of assemblages of relations between and within governments, political parties, corporations, nongovernmental organizations, municipal authorities, and the multiplicity of individuals and organizations that constitute everyday life.

It is then important to consider the question of what makes the relationships within the assemblage durable. Bruno Latour and Michel Callon have provided a way of conceptualizing the relationship between these disparate elements. They use the term "Leviathan" to describe the irreversible alliances created by the entanglement of human and nonhuman agents:

> In the state of nature, no one is strong enough to hold out against every coalition. But if you transform the state of nature, replacing unsettled alliances as much as you can with walls and written contracts, the ranks with uniforms and tattoos and reversible friendships with names and signs, then you will obtain a Leviathan: "His scales are his pride, shut up together as with a close seal. One is so near to another that no air can come between them. They are joined one to another; they stick together that they cannot be sundered" (Job 41:15–17).[30]

While in a relationship of cooperation with each other, the different entities are concurrently immersed in other assemblages of relations.[31] Susan Leigh Star and James Griesemer's concept of the "boundary object" offers a useful conceptualization of the meeting of interests that enable cooperation.[32] The boundary object straddles multiple worlds and is so able to bring heterogeneous elements together in a cooperative relationship of negotiation and compromise. In Star and Griesemer's case study of the Museum of Vertebrate Zoology at the University of California, Berkeley, each individual or organization involved collected and preserved flora and fauna in the state of California for preservation at the museum according to their own interests. In so doing, they retained their separate, individual identities while still cooperating with the projects of the museum's first director, Joseph Grinnell.

In the case of Iscor and Eskom presented here, conflict and compromise characterized the relationship between the state corporations and the apartheid government. The way forward for the state corporations was not always smooth, and intransigent opposition from affected ministries often blocked the paths they intended to follow. Their shared need to ensure the economic and military survival of White South Africa in the face of threats to apartheid from both within and without ensured their cooperation. This commitment to a defensive nationalism coincided with a faith in the efficacy of long-term planning. Eskom's and Iscor's long-term-demand forecasts of electricity and steel, respectively, determined the size of their ambitious expansion plans in the 1970s and 1980s, enabling their engineers' peculiar tenacity during this period. Brian Larkin has remarked on the evocative quality of infrastructure and its role in producing forms of fantasy and desire among those who interact with it that is independent of the "technical functioning" of an infrastructure.[33] As the story of Eskom and Iscor demonstrates, infrastructure can also evoke particular imaginaries of an ideal society and ensure cooperation and unity in the realization of this goal. These imaginaries could be recorded in government policy or exist outside of official government sanction. The existence of the Broederbond in twentieth-century South Africa is an example of the latter. The Broederbond was a shadowy organization responsible for the propagation of Afrikaner culture, education, and economic development, and it enjoyed a disproportionate amount of control over the National Party and the various leaders of the apartheid government.

FROM KEYNESIANISM TO NEOLIBERALISM

The apartheid government's tenacious commitment to building the infrastructure that would ensure its survival entailed significant government investment from the 1960s onward. This was in keeping with contemporaneous Keynesian global trends, where governments around the world hoped that public spending would invigorate ailing postwar economies. In addition to spending on infrastructure, by the 1980s the apartheid government was also spending large amounts of money on arms and artillery to repress ongoing protests in the townships and to prop up the treasuries of the homeland governments.[34] In 1985 South Africa officially entered a financial crisis, prompted by the refusal of the American-based Chase Manhattan Bank to extend further credit to the country. In response, the government turned its attention inward, and the state corporations swam into focus as targets for cost cutting. Thus began the serious consideration of privatizing the state corporations; a process that Ben Fine describes as "selling-off of the family silver in order to raise the funds to finance the apartheid regime."[35] President P. W. Botha instituted a committee to investigate the privatization and deregulation of state corporations, and the report that the committee released echoed neoliberal concerns about government overreach in the management of the economy and the importance of carving out a space for private sector involvement.[36]

Introduction ⁓ 11

This preoccupation with fiscal austerity and the privatization of state-owned corporations tailed similar developments elsewhere in the world and is generally indicative of the onset of neoliberalism. In European countries, neoliberalism had arrived slightly earlier. In the late 1960s, stagflation, or intolerably high inflation rates, stalled the Keynesian model of government spending to stimulate economic growth, and the proponents of neoliberalism, who had previously been sidelined, rushed to fill the now vacant space of economic orthodoxy. Neoliberals saw individual liberty as tied to market freedom and advocated the relaxation of the hold of governments in organizing the economy. On the African continent, neoliberalism is seen to have arrived with the structural adjustment policies (SAPs) imposed by the World Bank and International Monetary Fund on bankrupt governments as a part of loan conditionality. Because they advocated austerity, SAPs are thought to have destroyed governments' capacities to maintain the developmental projects introduced in the late colonial and postcolonial periods.[37] Nonetheless, as Nicolas van de Walle argues, it is difficult to establish the actual efficacy of SAPs, when measured in terms of their ability to realize their own objectives or in terms of the actual scale of the transformation they wrought.[38] South Africa, however, managed to escape adherence to the prescriptions of SAPs, though World Bank officials regularly advised both the government under apartheid and the government led by the African National Congress (ANC) on economic policy. The ANC inherited a decimated fiscus when it came into power in 1994 and implemented what is widely considered a neoliberal-inspired policy, Growth, Employment, and Redistribution.

While often presumed to be a foreign imposition on African governments, neoliberalism offered an opportunity to mend the fracturing apartheid project for government officials concerned with reform in the 1970s.[39] In South Africa, as elsewhere on the continent, the oil crisis of 1973 signaled the start of financial hardship, and African governments struggled thereafter to access the funding they previously could. In South Africa, the oil crisis was soon followed by the Soweto uprising of June 1976, which began persistent unrest in the townships and, as Deborah Posel writes, urged the apartheid government to adopt a "new language of legitimation."[40] Faced with the growing "ungovernability" of the townships, which was an avowed ANC resistance strategy, the National Party attempted to reform its strategy of racial segregation and subdue discord within its ranks. It failed, however, to prevent the eventual defection of a faction of the Afrikaner nationalist right in 1982 to form the Conservative Party.

Antina von Schnitzler writes that certain government economists found neoliberal precepts appealing and recommended that apartheid reform focus on molding the subjugated African in the image of the market consumer. Where it had previously eroded all forms of African capital ownership, in the 1980s, the apartheid government attempted to create an elite class of African entrepreneurs in the townships and the homelands in the hopes of alleviating popular protest.

The preoccupation with the commercial reform of the state corporations in the 1980s also reflected the South African government's suspicion that Eskom had too much free rein. By the end of the 1980s it became apparent that Eskom had built too much electricity generation capacity, and the concern with privatization coincided with the view among government officials and commentators in the press that public funds had been wasted on the new power stations. While Eskom escaped privatization in the 1980s, it came to the realization that its long-term planning of supply and demand had been misleading, inadvertently confirming the neoliberal argument that the market is best placed to determine levels of demand and supply. Eskom's demand projections proved erroneous because of the unexpected drop in gold prices in the 1980s, and this transformed the century-old alliance between energy in South Africa and the gold mines. At the same time, global and domestic forces agitating for the unbanning of political parties such as the ANC and for the implementation of democratic systems wore away the resolve of the apartheid regime.

While the neoliberal transition looms large in the history of the African continent, its implications and defining features have proven elusive. So significant is its apparent vacuity that Rajesh Venugopal has suggested we retain neoliberalism only as a "broad indicator of the historical turn in macro-political economy."[41] The difficulty in definition is due, in part, to the fact that its actual manifestation assumed unrecognizable forms. While attributed to the musings of members of the Mont Pelerin Society, who began to meet at a Swiss mountain resort in 1947, neoliberalism was always and everywhere grafted onto preexisting social fabrics, and this was part of its appeal for apartheid reformers.[42] For this reason, the preexisting institutional structures, nongovernmental organizations, political parties, infrastructure, and technological corporations shaped the adoption of neoliberalism around the world.

Scholars of neoliberalism have pointed to its ultimate role in exacerbating wealth inequality, while it paradoxically continues to rely on the regulatory authority of the state. Pierre Bourdieu has highlighted the

Introduction ⁓ 13

paradoxical role of neoliberalism, which criticizes "collective structures" that oppose its valorizing of the individual, while these older structures remain responsible for curbing the social chaos that neoliberalism would unleash if left to its own devices.[43] David Harvey describes the "neoliberal state" as one that is committed to the accumulation of wealth for a multinational business elite.[44] Efforts to suppress the economic mobility for the poorest of society follow, such as the erosion of welfare and the relaxation of laws that protect labor. Scholars have also considered the continued provision of government welfare through the use of neoliberal techniques, such as in James Ferguson's study of direct cash transfers in the rollout of South Africa's Basic Income Grant.[45] Similarly, Stephen Collier reveals that state corporations continued to assume responsibility for the provision of heating in post-Soviet Russia, all the while utilizing elements of the neoliberal toolkit, such as privatizing parts of themselves, an exercise known as "unbundling."[46]

Neoliberalism is also considered to have a splintering effect on the large-scale, networked infrastructure, much like the large technical systems that Hughes has described, common under a state-centered model of development.[47] Privatization and the effort to encourage competition entailed the breaking up of these behemoths at both the national and local level, leading, in cities, to what Stephen Graham and Simon Marvin have termed "splintering urbanism."[48] Where new technologies have been adopted, these often take the form of "micrological" devices, such as the water and electricity prepaid meter that Von Schnitzler has described, in line with devolved, individualized techniques of neoliberal governmentality.[49] But the tendency toward the splintering of infrastructure has existed side by side with an impetus toward centralization. The persistence of large, networked infrastructures in a nominally democratic South Africa with a competitive market economy is best explained by the continued advocacy—typically by trade unions in official forums—of "democratic socialism." Eskom and the South African government repeatedly toyed with the idea of the privatization of Eskom in the 1990s, but it remained a potent state corporation with the ability to build enormous plants such as the 4800 MW Medupi power station. While not disputing its splintering effect, the implementation of neoliberalism is viewed here as a contested process. In South Africa, the governmental preoccupation with privatization and fiscal austerity, activities associated with a neoliberal orthodoxy, has ebbed and flowed in a cyclical fashion since the 1980s. Trade unions in particular have been strong proponents of "democratic

14 ~ *Apartheid's Leviathan*

socialism" and have acted as a countervailing force to the imposition of neoliberal-inspired policies. They have done this from a position of power as a part of the ruling "tricameral alliance" that is composed of the Congress of South African Trade Unions, the South African Communist Party, and the ANC.

DEMOCRATIZATION

Across the African continent in the early 1990s, democratization followed on the heels of economic liberalization.[50] Frustrated at the slow pace of economic reform, international monetary bodies urged the implementation of multiparty elections in the belief that authoritarian governments were stifling economic growth. Democratization, culminating in South Africa's multiparty elections of 1994, institutionalized party contestation. Multiparty elections, while holding the promise of freedom and liberation from authoritarian rule, also brought new forms of insecurity and uncertainty. This was the case in Rwanda, where genocide began in the same month as South Africa's democratic elections.[51] The contestation formalized in the democratic process also meant the splintering of the imaginary that had previously animated the relationship between the government and the state corporations. As a crucible of politics in postapartheid South Africa, an infrastructure project such as Medupi became capacious enough to hold multiple contested assemblages, composed of both the human and the nonhuman. The infrastructure project absorbed a multiplicity of expectations and imaginaries of the future by virtue of its complexity and its extended time duration.[52]

The Medupi power station promised to end load shedding, periods of forced electricity outages that have afflicted South Africa since 2007. But the construction of the power station was ultimately uncontrollable. Its technological ambitions proved expensive and extremely difficult, and its continued contribution to climate change became indefensible. Medupi originated within the milieu of a resurgent focus on government spending on infrastructure to encourage a Keynesian-like stimulus of economic growth. Increased government spending was a common response around the world to the depression wrought by the financial crisis of 2007–8.[53] In South Africa, this coincided with the new presidency of Jacob Zuma, who was elected at the end of 2008. Zuma rapidly dissolved any pretense at austerity, and government funds lined the pockets of politicians and businesspeople as much as they went toward the construction of infrastructure. But this period did not exactly mirror the events that occurred during the

Introduction ∽ 15

period of Keynesian-inspired government spending from the Second World War to about 1980. In particular, the animating logic of the ruling party had changed. The National Party had focused on constructing the infrastructure of racial segregation as well as the cultural and economic promotion of Afrikaans and of Afrikaner unity encapsulated in its idea of the *volkseenheid.* But for the ANC-led government, the need for party funding underpinned its corruption scandals (as well as those of opposition parties) in the context of a party with a history of militant struggle that lacked inroads into the country's higher economic echelons before it came into power.

In a certain respect, the transition from apartheid to democracy with regards to infrastructure has meant a transition from engineers' tenacious conviction in the rectitude of their activities to contestation and irresolution. Scholars writing on the way in which questions of science and infrastructure are imbricated in democratic politics have highlighted the importance, and desirability even, of controversies.[54] Technological controversy benefits the practice of democracy because it allows ordinary citizens and consumers to inform the direction of technological change, removing decision-making from being the sole province of scientific and technical experts. This creates new forums for democratic action and consultation and ensures the participation of affected parties, through which the natural and social orders are "coproduced."[55] In this way, uncertainty enables participation by people who would otherwise be marginalized in decision-making about the direction of technological change.

The case under discussion here offers a layer of complexity to current understandings of the implications of democracy for technology and infrastructure. In principle, the ANC was elected into power as the representative of the electorate, but its leaders are also responsible for ensuring the survival of the party (which is the representative of the people). The party has profited off its control of the levers of the government, and since this behavior is largely illicit, the line between corruption for the party and for individual politicians is easily crossed. In the case of Medupi, uncertainty over the technical diagnosis of its problems acted as a cover for competent and well-meaning leaders in Eskom to be removed in favor of those who sought to siphon funds for individual enrichment. As a result, uncertainty was not resolved in a manner that enabled a consensus among affected parties and was instead used as an instrument to enable looting to continue unabated. A complex megaproject such as Medupi is the site of multiple, overlapping, and changing alliances, which made it difficult to identify a single entity or individual that is responsible for the project's failures. The long-standing nature of the construction and the fact that costs were escalated in a seemingly unlimited manner meant that a construction project such as Medupi was suitable terrain to absorb the pressures of democratic politics in South Africa.

METHODOLOGY

This book uses primary sources from various archives. These include the National Archives of South Africa, which houses much of Iscor's archive from the 1970s. I also utilized Eskom's own archival documents. These were obtained with special permission which specified that Eskom had the right to read the chapters written for my dissertation (on which this book is based) that mention the information contained within its archival records. I duly submitted chapters 3, 4, and 5 of the dissertation for their perusal, and Eskom requested no change apart from changing the reference to its earlier name of "Escom" to "Eskom." I obtained some archival material from certain interviewees who belonged to trade unions at the Matimba power station. These interviewees had kept the minutes of meetings between management and trade union representatives held at the power station in the 1990s and were helpful in reconstructing the narrative of negotiations that occurred during this period. I also utilized government publications to gauge the official parliamentary view of pertinent historical events. These include the records of parliamentary debates (Hansard), various white papers, and the reports of commissions of enquiry set up by the government.

When I began research on Medupi in 2013, Eskom had announced the first postponement of Medupi's completion date, which was initially the end of 2013. I approached Medupi with the intent to locate the points at which the autonomy of Eskom and its engineers had been eroded, rendering Eskom subject to political interference that sacrificed technical efficiency. But the story turned out to be more complicated. I decided that it would be unwise to interview engineers or staff members at the power station because of interviewees' likely guarded responses. In addition, the fact that Medupi was an ongoing construction project meant that it would have been difficult to gain a sense of developments there from any isolated section of interviewees. Casual conversations revealed that there were as many different points of view for the reasons for Medupi's failures as people I spoke to. For this reason, I have relied chiefly on documents in the public domain—news articles and reports of various Commissions of Inquiry—to establish the nature of events at the power station. Evidence given to the Commission of Inquiry into Allegations of State Capture (established in 2018

Introduction ⁓ 17

and also known as the state capture commission) is discussed in relation to the travails of the Medupi power station, though it is important to note that these witness testimonies are delimited by the parameters of the inquiry. I conducted interviews, chiefly as life histories, for the earlier period of the book because I believed that the passage of time would render the relation of past events less controversial. A few long-standing residents of the Lephalale and of Marapong were helpful and happy to share their memories. I also conducted interviews with engineers who had worked at Iscor and Eskom during the 1970s, 1980s and 1990s. Efforts were made to cross-check their information with archival documents as much as possible.

The archival documents I consulted for Iscor and Eskom were chiefly the minutes of board meetings that were held roughly each month in the 1970s and 1980s. Iscor was privatized in the late 1980s, and its company archives before privatization have been deposited in the National Archives of South Africa and are freely accessible to the public. The documents for the relevant years related to Eskom's activity are housed in Eskom's internal archives. The minutes of the board meetings for both Iscor and Eskom revealed the decision-making process that led to them entering the Waterberg in the mid-1970s. Since these were official records, they relay the impression of an eminently rational decision-making process, one in which the costs and benefits of all possibilities were considered in order to reach the best possible solution. This apparently rational process was itself an artifice since it depended on the reduction of complex economic, social, and political factors into factors that could be manipulated in a cost-benefit model. For example, the records contain no mention of the international and domestic hostility toward the apartheid regime during this period. The record of Eskom board meetings refers to the threat of the underground struggle launched from ANC bases in other African countries only as "defense" considerations. The tale relayed in this book relies on these documents, while euphemistic in the extreme, to understand the relationship between Iscor, Eskom, and the government as well as their official motivations for entering the Waterberg.

Unlike Iscor, Eskom was not privatized and remains a state-owned corporation at the time of writing. Many of the engineers who rose through its ranks in the 1970s and 1980s were still employed at the corporation during the course of my research. By the mid-2010s, allegations of government interference in tendering processes had become clear, raising concomitant fears that this would threaten the technical efficiency of existing and newly commissioned power stations. As the years wore on, the scale of the penetration

18 ⁓ *Apartheid's Leviathan*

of state capture operatives became more evident and engineers who held management positions at Eskom testified at the state capture commission. These engineers revealed a sense of incomprehensibility about the governmental interventions, which had manifested in sudden dismissals of personnel for reasons that were unconvincing. While Eskom's engineers did not launch overt protests, there was a sense of frustration at the government's interference in internal decision-making processes. There was also a sense that events were occurring that were outside of the control of Eskom's engineers. The controversies surrounding the state capture scandal touched on events at Medupi, and the ANC had been implicated in the improper award of the tender for the boilers early in the construction of the power station. While an investigation by the public protector of South Africa in 2008 laid to rest claims of ANC interference in the award of the tender, there remained a veil of suspicion around how corruption at Medupi was influencing the continuous delay in completion of the power station.

NOTES ON TERMINOLOGY

I have chosen to use the names of places that are contemporaneous with the period under discussion. For example, the name of the town of Ellisras was changed to Lephalale in 2002, but when discussing the history of the town during the 1970s, the name Ellisras is used in accordance with the terms of the discussion in archival records. The exception to this is the name "Eskom," which was known as "Escom" before 1987.

Since racial segregation is an important component of South Africa's history, the use of racial categories is unavoidable in a study such as this. During the antiapartheid struggle, the term "Black" came to encompass the so-called racial groups that bore the brunt of apartheid's discriminatory laws. These groups included the apartheid-created racial categories of "Black," "Indian," and "Coloured." With the absence of the racial solidarity occasioned by the antiapartheid struggle, in the postapartheid period it is difficult to ascribe the same cohesion of racial categories to the term "Black." I have used the term "African" to refer to indigenous South African peoples, whom the apartheid government classified as "Black." The term "Black" is used in the book to denote the racial groups that the apartheid government classified as "Black," "Indian," and "Coloured," as described above.

THE INTERNAL ARCHITECTURE OF THE BOOK

Chapter 1 details Iscor's arrival in the Waterberg. Located far from the infrastructure of coal exploitation, the Waterberg was an unlikely site of state

Introduction ⌁ 19

corporation activity. During the 1960s, the apartheid government worked closely with Iscor and Eskom to realize its project of national industrialization and racial segregation. As a frontier-like border region without a substantial White settler population, and surrounded by fragments of the self-governing Lebowa homeland, the Waterberg was not of any particular importance to the government. Iscor drove the exploitation of the Waterberg coalfields to meet the coking coal requirements of its expansion plan, which was in turn determined by its demand forecasts. Despite the threat posed by the global scarcity of funds after the oil crisis of 1973, Iscor proceeded tenaciously with the development of the Grootegeluk coal mine in the Waterberg. Iscor and the government came together to enable the unlikely exploitation of the Waterberg coalfields, united by a shared nation-building project in a relationship characterized by conflict and compromise.

Chapter 2 focuses on the development of the small town of Ellisras and the mediated way in which it was subject to the government's regulatory authority. As a vast expanse of bushveld, the Waterberg was too far from the government capital, in Pretoria, to feel the full might of governmental control. It was only with Iscor's arrival in the mid-1970s that the government turned its eye to regulating urban development and racial segregation in the incipient town of Ellisras. Iscor's arrival, with its promise of large-scale capital investment and urban growth, coincided with the forced removals of Africans from White-owned farmlands in the district in accordance with the Group Areas Act. This enactment of forced removals, a characteristic feature of authoritarian high modernism, was not directly a product of governmental decree. While the Group Areas Act was one of the pillars of apartheid, the government lacked the will or the wherewithal to commence with forced removals everywhere in the country. Iscor provided the infrastructural muscle to ensure the development of a "modern" town, and in line with government prescriptions, a modern town was also one that was racially segregated. The forced removals of African communities from the vicinity of the town to the nearby homeland was enabled by a confluence of concerns from Iscor and various layers of government, including labor scarcity, public health, town planning, and racial segregation.

Chapter 3 details Eskom's arrival in the Waterberg in the 1980s as the apartheid regime increased its military capacity in defiance of its impending end. The Matimba power station, built near Iscor's coal mine in the Waterberg, was a part of Eskom's major power station construction spree in the 1980s. A power station in the Waterberg required significant technological innovation to cope with the arid climate, and its construction is a testament

20 ～ *Apartheid's Leviathan*

to Eskom's tenacious commitment to its own capacity expansion plan. At the end of the 1980s, however, Eskom's erstwhile foundation of certainty began to crumble. In the end, it had created too much electricity generation capacity. The collapse in the price of gold meant that demand from the gold mines did not increase as much as Eskom had predicted, belying the accuracy of long-term planning. Eskom also encountered rare opposition to its plans in the form of the country's air pollution officer, who forced the corporation to situate one of its power stations outside of its traditional stronghold of Mpumalanga, which had become saturated with sulfur dioxide emissions.

Chapter 4 details the beginning of the neoliberal era in the 1970s. Following the Soweto uprising, apartheid underwent a process of reform that saw the political triumph of the *verligte* faction, a group that advocated an embrace of commercial principles even if this meant the relaxation of racial segregation. The rising tide of neoliberal orthodoxy offered a means of commercial salvation, and when the government entered a financial crisis in the late 1980s, government officials explored the option of privatizing state corporations to gain liquidity. This chapter demonstrates the selective and partial incorporation of neoliberalism as a top-down attempt at reform. The privatization of Eskom ultimately appeared too inconvenient. The corporation had proven its role in diplomacy in the southern African region under apartheid, and for the incoming ANC-led government, it promised to deliver universal electrification and the fruits of modernity to the previously disenfranchised citizenry. In this way, Eskom resisted privatization and total divorce from the levers of the government while still committing to commercial reform.

Chapter 5 details the arrival of African trade unions at the Matimba power station and their incorporation into official labor bargaining forums. Labor organization at the Matimba power station and at the nearby Grootegeluk coal mine followed a coal mine and power station trajectory of union organization that occurred elsewhere in the country, particularly in the Gauteng-based industrial hub of the Vaal Triangle. Because of the proximity of power stations to coal mines across the countryside, the National Union of Mineworkers and the National Union of Metalworkers of South Africa were naturally inclined to organize in a similar way in the Waterberg. Trade unions negotiated the transition from paternalism to the idea of workers as nominally autonomous individuals, highlighting the complexity of the transition in a context of deep dispossession. With the award of South African citizenship rights for Africans in the town of Ellisras after 1994, the

Introduction ～ 21

power station became an important site for the promotion of autochthony, as workers across racial lines agitated for residents of the region to be prioritized for employment and promotion opportunities.

Chapter 6 details the development of the Medupi power station, demonstrating that the power station functioned as an entity capacious enough to absorb shifting alliances and imaginaries in the democratic era. South Africa adopted a resurgent infrastructural drive toward the end of the first decade of the twenty-first century, one that shared features with the Keynesian economic stimulus after the Second World War. Despite the best political will—given the public discontent at the frequency of electricity outages and its importance to the ANC's electoral prospects—Medupi consistently defied estimates of the dates of completion of construction. Eskom promised at the outset to install pollutant reducing technology, called flue gas desulfurization, at Medupi, though it has delayed doing so. Medupi is an air polluter and a contributor to the ever-worsening climate change crisis. Over time, state capture operatives, or those who sought to illicitly profit from the power station construction, manipulated the facts behind its delay as an excuse to remove competent engineers from Eskom when, in reality, no single individual could be held responsible for its lack of completion. In time, irresolution and uncertainty over the power station's failures were used as a cover for continued looting. This allowed state capture operatives to plant pliable officials in the managerial ranks of Eskom—these officials assisted in the project of looting and were not particularly concerned with the maintenance of infrastructure. Medupi and Eskom have ultimately drawn the government into a position of indebtedness from which it cannot easily escape. Eskom now stands out as a globally critical contributor to the climate crisis and a major threat to the government's fiscal well-being. Whether Eskom remains a state corporation or is wholly privatized, it is not likely to construct another coal-fired power station of a similar scale in the near future.

This book sets out to illuminate the infrastructural, technological, and material dimension of politics in South Africa and on the African continent more generally. Politics in South Africa is generally considered to be people-centered—its trajectory dependent on the machinations of political parties, influential politicians, and citizens at the voting booth. The corpus of science and technology studies, with its focus on the vitality of the material, has complicated the notion of intentionality, highlighting the intermediate elements between command and action. During the 1960s and early

1970s, the period of apartheid rule that closely approximates James Scott's notion of authoritarian high modernism, the institutional autonomy of the technological state corporations, the complexity of the technology, and the geography of the country's mineral deposits, to name a few, complicated the apartheid government's ability to enact its will. The technological projects of the Waterberg originated in this foundational period of government-driven infrastructural development, and an examination of their evolution reveals the contingent, inadvertent nature of authoritarian high modernism.

This contingency extended to the project of extending the authority of the government to peripheral regions of the country, which affected the apartheid government's ability to effectively control the people and things within the territory of the nation-state. An important theme in studies of the African state is that of its difficulty in gaining legitimacy in the eyes of its populace and effective control over the territory within its borders. But infrastructure and technology were not passive transmitters of government power, and their intermediary role means that the transmission of government power was the product of an assemblage of factors composed of various layers of government officials, engineers, local elites, labor, and materials. This challenges the presumption of concerted action contained within the notion of the "African state."

The institutional autonomy of the state corporations, Eskom and Iscor, meant that they did not perfectly align with any presumed role that politicians envisioned for them to play. In this way they played an ambivalent role—or contradictory roles simultaneously—in South Africa's historical development. For example, analysts of South African political economy consider Eskom to have played a key role in sustaining the particular set of capitalist relations contained within the minerals-energy complex. But at the same time, Eskom was also important to the government's developmental project—focused on the improvement of the living conditions of Whites—for most of the twentieth century. Iscor and Eskom merged with the imperatives of successive governments of the twentieth and twenty-first centuries and so passed through different technopolitical regimes. This includes the austerity of the neoliberal era and later the renewed infrastructural emphasis of the ANC-led government. While a part of these regimes, they were also transformed during the course of their passage, incorporating new elements and retaining traces of their past activities.

The major transition of South Africa's history—from the oppressive rule of apartheid to the freedom of democracy—is complicated by the ambivalent presence of large technological systems in the Waterberg, and more

Introduction ⁓ 23

generally by the networked infrastructure in the form of Eskom, its country-wide network of power stations, and its extensive network of transmission lines. The technological systems in the Waterberg, which were created in the midst of apartheid, set the stage for the formation of African worker solidarity in the region. In the postapartheid era, the construction of the Medupi power station in the region has provided hope for the end of the country's electricity shortage crisis and for the delivery of the fruits of modernization to the majority of the populace. At the same time, the delayed completion of Medupi has frustrated efforts to realize the material benefits of democratic freedom, for which access to electricity is crucial. The technological systems in the Waterberg have come to signify both freedom and unfreedom from their origins in the apartheid period and in their continued activity in the postapartheid era.

In understandings of the political economy of the transition from authoritarian to democratic rule in Africa, neoliberalism immediately preceded the democratic turn. The oil crisis of 1973 was a crucial moment in ending the overbearing presence of African governments in the economy. But the relationship between the global funding scarcity, signaled by the oil crisis, and the end of the government-led developmental project is a complicated one. The development of the technological systems in the Waterberg continued into the 1980s despite the funding scarcity. In accordance with the argument made by scholars such as Stephen Collier, the neoliberal era did not automatically mean the privatization of state-owned corporations such as Eskom. This book further develops this point, demonstrating that the imposition of neoliberalism was a contested process. Trade union organization in the Waterberg in the 1980s closely followed the organizational experience among the coal mine and power station nexus of the Vaal Triangle, in what is today part of the Gauteng province. In this way, the trade unions negotiated and contested the imposition of principles associated with neoliberalism in a context where Africans had been steadily dispossessed of capital over the course of the twentieth century. Neoliberalism was thus imperfectly adopted and the shape it assumed was a product of preexisting material and organizational configurations such as those set by the coal mine and power station nexus of the Waterberg.

South Africa's infrastructure networks have stealthily gained visibility in the postapartheid era due to innumerable instances of failure, in line with Paul Edwards's contention that infrastructure is largely invisible until breakdown. Electricity provision in particular has come to be a crucial measure of the health of the country's democracy and of the satisfaction of the general populace with the rule of the ANC-led government. Much like other state corporations, Eskom has fallen prey to corruption, which has resulted in the widespread hollowing out of infrastructure and the organizations that sustain them. The construction of the Medupi power station has proceeded through three different presidential regimes, and it has proven too complex to relay a single diagnosis that could determine an appropriate remedy. Repair and correction have occurred during the course of construction in a feedback loop of sorts, leading to the continual delay in completion. Medupi and the technological systems in the Waterberg are both subject to, and immersed in, the economic, financial, technological, and political milieu of democratic South Africa. In this way, they have proven crucial to the country's prosperity.

Introduction ～ 25

1 ～ The Unlikely Exploitation of the Waterberg

ON 21 March 1960, police officers opened fire on a group of unarmed protesters in Sharpeville, a township to the south of Johannesburg, killing 69 people and injuring 130. The protest had been organized by the Pan Africanist Congress (PAC) to protest the pass laws that the apartheid government had introduced to control and monitor the presence of Africans in urban areas. After the Sharpeville Massacre, as it became known, the apartheid government intensified its repression and banned the African National Congress (ANC) and the PAC, both of which launched underground struggles in response. The Sharpeville Massacre laid bare the atrocities of the apartheid regime for the world to see, and South Africa consequently faced the very real threat of international sanctions. But rather than compromise its commitment to racial segregation, the government instead bolstered its security apparatus. This began under the leadership of Prime Minister Verwoerd, who was assassinated in 1966. In 1969 his successor, Prime Minister John Vorster, formed the Bureau of State Security, a police structure that later formed covert death squads responsible for killing those with suspected links to the ANC in exile.

As the government oiled its machine of terror, it also battened down the hatches and stepped up the pace of industrialization. This would ensure the country's self-sufficiency in the likely event that the international community imposed sanctions against South Africa. But private sector investment had dissipated in the early 1960s due to two political shocks—the Sharpeville Massacre and South Africa's withdrawal from the Commonwealth of Nations in 1961.[1] To stimulate economic activity, the government increased public sector investment to its highest level of the twentieth century, between the midsixties and 1976. Fixed public investment increased from 45 percent in 1963 to 53 percent in 1976, before declining to 27 percent by 1991.[2] State corporation spending made up a large portion of this public sector investment because the state corporations provided affordable inputs, such as steel and electricity, for manufacturers. The project of government-driven industrialization soon bore fruit and in 1962, members of Parliament praised the government for investing in Eskom and Iscor and acknowledged the role those state corporations played in stimulating commercial activity in the private sector.[3]

The government also saw state corporations as a peg in the grand scheme of regional development, which envisioned economic growth and scientific progress to play a role in preventing the depopulation of the countryside by ensuring that White residents of rural districts had sustainable livelihoods.[4] In 1966, Prime Minister Hendrik Verwoerd announced the establishment of a new department, the Department of Planning and Coordination, which had identified fifteen regions that could benefit from government-coordinated development.[5]

African homelands, or the bantustans, featured in this conception of the "regional" as pools of cheap labor. While the apartheid government envisioned the homelands as becoming autonomous, self-governing nations, their isolation from the White-dominated centers of wealth threw their economic viability into question. The borders of the homelands were thus strategically drawn so that they were adjacent to nominally White towns and could serve as pools of cheap labor. From the late 1960s, through this policy of "industrial decentralization," the government encouraged corporations to situate their factories in the industrial districts of these small towns. Incentives included direct government subsidies and the appeal of the cheaper labor of the rural areas as opposed to the more militant urban workforce.[6]

Iscor and Eskom created huge plants which, much like the large technical systems that Thomas Hughes has described, relied on economies of scale, utilized a centralized network, and displayed an organizational complexity that blurred the managerial line between the social and technical.[7] Since these plants were situated on large tracts of land in undeveloped areas, they promised both to increase economic and infrastructural development and to prevent the depopulation of the countryside. Iscor also owned some of

The Unlikely Exploitation of the Waterberg ⟿ 27

the mines that produced its mineral supply and, as a result, created a geography of steel based on the country's iron ore and coking coal reserves. Coking coal, which is of a higher quality than the bituminous coal commonly used in household fires, acts as a reducing agent to transform naturally occurring iron oxide (Fe_2O_3) to crude iron (Fe) during the smelting process. Crude iron is then further processed with carbon to form steel. Precolonial iron smelters commonly used wood as fuel during the smelting process.[8] Over time, coking coal became the accepted fuel for iron smelting because it did not release as much smoke as lower-grade bituminous coal. During the course of the twentieth century, Iscor bought over (or started from scratch) iron ore and coking coal mines that were connected by railway lines to ports or steel plants in a network that traversed the South African countryside. By 1970, Iscor owned the Sishen iron ore mine in the deserts of the Northern Cape, the Thabazimbi iron ore mine in the Waterberg, and the Durban Navigation Colliery, set among the rapidly declining coking coal reserves in KwaZulu-Natal. Because of the scarcity of coking coal, Iscor had to protect its domestic supply from being exported during cyclical global booms. In the aftermath of the oil crisis of 1973, global demand for South African coal grew and coal producers rushed to meet this demand. The imperative of industrialization for Iscor, combined with the opening of global markets for South African coal, ultimately led to the unlikely exploitation of the Waterberg coalfields. The Waterberg was the final coal frontier in the 1970s, and its exploitation became the subject of intense negotiations between Iscor, the government, and Eskom.

COAL AND IRON IN SOUTH AFRICA'S INDUSTRIAL REVOLUTION

Jan Smuts, who served as prime minister of South Africa from 1919 to 1924, and again from 1939 to 1948, encouraged the creation of the state corporations to encourage industrial development and preserve the color bar. The color bar entailed job reservation in accordance with racial classification and, for the mines, meant that White laborers performed skilled work while African laborers were restricted to lower-skilled and lower-paid work. The gold mines were the most important economic activity in the country in the early twentieth century and so shaped the nature of labor relations in South Africa. When gold mine owners were left to their own devices, they searched unceasingly for lower wages, which threatened the sustainability of the color bar and the project of White economic supremacy. In June 1913 Smuts, who served as minister of mining at the time, brutally suppressed a strike by White mineworkers, who were incensed by the threat that lower-paid African laborers posed to their job security. Fully aware that the White

28 ~ *Apartheid's Leviathan*

working class was a ticking time bomb, Smuts subsequently saw the state corporations as a means to support mechanization in the mines and in the manufacturing sector.[9] Mechanization cemented the color bar because it meant that corporations could utilize a mass of low-skilled African labor and skilled White labor operatives. The railways, overseen by South African Railways and Harbours, provided important opportunities for Whites, and Smuts hoped that Eskom would supply inexpensive electricity to the railways despite the mines' protestations that they would be subsidizing the cost of running the railways through the price they would pay for electricity.

The rise of Iscor and Eskom signified the coalescing of coal and iron in South Africa's industrial revolution. Their progenitor, Hendrik Johannes van der Bijl, came from a long line of Dutch descendants in South Africa. He had left South Africa in 1909 and went on to earn a doctorate in physics from the University of Leipzig. Van der Bijl was an internationally renowned scientist, working as a nuclear physicist in New York on the Manhattan Project, when Smuts urged him to return to South Africa to lead the development of the state corporations.[10] In an interview with the *Rand Daily Mail* shortly after his appointment, he espoused the virtues of coal and iron and subsequently prioritized the development of the steel industry and the provision of cheap electricity to the country. In 1922 he expressed his fears about South Africa's sole dependence on gold, which he termed a "wasting asset."[11] Iscor was formed in 1928 in what Bill Freund has described as a "daring act" by the Pact government, because the mining industry opposed the creation of a government-controlled steel manufacturer.[12] Van der Bijl brought in teams of foreign contractors to help with setting up the steel production process and to train South African engineers.[13] Regardless of the nature of Smuts's purposes in setting up the state corporations, van der Bijl declared that his role lay outside the sphere of government. His was a closely guarded autonomy. In 1939 he said, "At present I have no enemies that I know of, but if I join the Cabinet I shall immediately have 40 percent of the population against me and I shall have to waste my time making conciliatory and tactful speeches."[14]

A long history of iron smelting in southern Africa preceded the creation of Iscor. Since the advent of the Iron Age, which archaeologists have dated to the immigration of African agriculturalists in AD 200, metalworking has been important to southern African societies. A wealth of evidence demonstrates that precolonial iron smelting took place in the Waterberg, particularly in the surrounds of the town of Thabazimbi, which lies about eighty miles (130 kilometers) south of Lephalale.[15] In November 1961, at a presidential address of the Associated Scientific and Technical Societies of

The Unlikely Exploitation of the Waterberg ～ 29

South Africa, a member of Iscor, Dr. F. Meyer, described what he termed the impressive, "primitive" steel-making traditions of Natal. By his account, British colonial official Theophilus Shepstone had looked on with awe at the workings of a Zulu ironsmith whom he encountered in 1853. Shepstone took a piece of the iron with him and showed it to a European blacksmith in the town of Pietermaritzburg, who described it as "superior to the 'best English iron' and equal to the Swedish iron."[16] In the Transvaal, the gold mines and surrounding industry created a sustained demand for steel, and some entrepreneurs tried their hands at steel manufacturing. The most notable of these was Sammy Marks, who had initiated commercial agricultural activity on land along the banks of the Vaal River and acquired the Vaal River Colliery. Marks was born in Lithuania, where he gained some familiarity with steel making, and in 1911 he created the Union Steel Corporation of South Africa.[17] By the 1920s, three steel factories were in existence in the Transvaal.

After its formation, Iscor struggled to gain a foothold in a South African market dominated by imported European steel. To make matters worse, European steel manufacturers were "dumping" their steel in South Africa, which meant that they exported the steel that they could not otherwise sell to South Africa's shores, where it undercut locally manufactured steel and so destroyed the profitability of the latter. Iscor's fortunes only began to improve with the onset of the Second World War when, preoccupied by warfare, European economies turned inward and South African demand for domestically produced steel rose. Iscor thus emerged from the Second World War in a much stronger position, having gained monopoly control of the South African steel market. Soon after the end of the war, in 1948, the National Party rose to power with its promise to implement apartheid. Hendrik van der Bijl passed away in the same year and his prodigal successor, Hendrik van Eck, took up the leadership mantle. Van Eck, like van der Bijl, was a graduate of the University of Leipzig and was well attuned to van der Bijl's technocratic vision of harnessing the country's scientific prowess to ensure its self-sufficiency. From 1942 to 1952, he had served as the chairman of the Social and Economic Planning Council, an entity that Smuts's wartime government established to advise the government on developing economic policy and on managing national resources. In an article published by the *Journal of the Royal African Society* in 1942, Smuts praised the work of the Social and Economic Planning Council and urged it to "avoid purely political considerations," stating that "you can but do your best in an impartial, objective manner, following as much as possible scientific lines."[18]

The Social and Economic Planning Council released a report entitled *Public Works Programme and Policy* in 1946, which deployed Keynesian concepts that were becoming globally prominent in the aftermath of the Second World War. The council emphasized the importance of government spending to stimulate the postwar South African economy and imagined that this spending would be channeled through government-driven construction projects, or "public works." Citing the report of the British Poor Law Commission of 1909, it argued that public work programs would not only create employment during construction but would also stimulate activity in the private sector.[19] Government spending would create a ripple effect in the economy, with an estimated multiplier ratio of one to three.[20] The council saw the Tennessee Valley Authority (TVA) in the United States, which was created in the 1930s and consists of a series of hydroelectric dams built along the banks of Tennessee River, as a model of coordinated regional planning. In one fell swoop, the TVA had controlled the flooding of the river and enabled soil conservation measures, rural electrification, and industrialization. As Albert Hirschman writes, agents of development projects all over the world looked upon the TVA with awe and fascination. Hirschman argues that the evocation of the TVA as a model to be emulated helped create the bravado necessary for governments and contractors to commence with the risky undertakings that are large infrastructure projects.[21]

Hendrik van Eck, armed with a technological vision of national development, drove the activities of the state corporations from 1948. Prime Minister D. F. Malan, leader of the newly installed National Party (NP), mistrusted the institutions of scientific research, such as the Centre for Scientific and Industrial Research, and the various state corporations.[22] For the Afrikaner nationalists these institutions bore the indelible imprint of former prime minister Jan Smuts and his British imperial sympathies. But the relationship between the NP and the state corporations improved with the turnover in the leadership of the NP. This was partly a matter of personality. Hendrik Verwoerd, known as the architect of grand apartheid, became prime minister in 1958, and he lacked the religious adherence of one of his predecessors, Malan.[23] Verwoerd had an established academic career before he entered politics: he held a doctorate in psychology and served as a professor of sociology and social work at the University of Stellenbosch. During his term as prime minister, Verwoerd embraced the power of science and technology to enact his ambitious projects of social engineering. He enjoyed a close relationship with the leadership of the state corporations, and Saul Dubow writes that he was rumored to have had a direct telephone line to van Eck's office at one of these, the Industrial Development Corporation.[24]

The Unlikely Exploitation of the Waterberg ⁓ 31

The oil crisis of 1973 only intensified the apartheid government's determination to ensure its self-sufficiency by building its technological expertise. Energy security took center stage in parliamentary debates at the time. The records of parliamentary debates in 1974 reveal parliamentarians' fears of South Africa's creeping isolation, encouraged by rumors that an alliance was brewing between the Middle Eastern oil-producing states and certain African countries within the Organisation of African Unity. Hostility to the apartheid regime threatened South Africa's supply of imported oil, and opposition members of Parliament blamed apartheid's racist policies for the animosity. In response, the minister of economic affairs reassured them that South Africa's nuclear prowess would enable it to retain a position of strength despite the unpopularity of the apartheid regime: "With South Africa's vast uranium resources and advanced techniques, we are in a position to play a very important role in the development and establishment of new energy resources for the world, and let me assure you that the outside world is aware of this," he said. "Our advanced development has enabled us to co-operate with and grant assistance to numerous states in the technical and scientific spheres, states not only in Africa but also in Latin America and the Middle East."[25] In this vision of South Africa's place in the world, technical and economic self-sufficiency would buffer the country against international hostility while ensuring the longevity of the apartheid regime.

ISCOR'S EXPANSION PLAN

By the 1960s Iscor was in complete control of South Africa's steel market, and in 1968 it announced an expansion plan on the basis of a predicted increase in demand for steel. This meant the construction of new steel plants, and Iscor worked in close consultation with the government in deciding where these would be situated. One of the new plants was built in the town of Newcastle, in what is today known as the KwaZulu-Natal province. Newcastle was then a government-designated "border industry," adjacent to the KwaZulu homeland. Iscor's steel plant boosted the economic fortunes of the town, rendering it an industrial hub that employed artisans and developed artisanal skills. When the minister of economic affairs announced in Parliament in 1969 that Iscor's new steel plant would be situated in Newcastle, he stated that in choosing the site, attention had been paid to the "wider social and developmental requirements within the framework of the government's policy."[26] The presence of an established White community in the town was an important part of its appeal, as were other factors, such as the availability of transport infrastructure, water, electricity, labor, and the long-term future of steel exports.

32 ～ *Apartheid's Leviathan*

Iscor hoped not only to produce steel for the South African market but also to flex its muscles on the global market. In the 1960s Japan had grown to become a global economic powerhouse and, with its industrial engine running, exhibited a near-insatiable appetite for raw materials. In anticipation of selling to Japan, iron ore mines in Australia and South Africa geared their production to meet its demand.[27] Iscor proceeded with great energy and built the 861-kilometer Sishen–Saldanha railway line (which opened in 1976), intruding on the domain of South African Railways and Harbours, which ordinarily oversaw the construction of railway lines. The railway line stretched across the Karoo desert from the Iscor-owned Sishen iron ore mine, located in the Northern Cape town of Kathu, to Saldanha Bay, in the Western Cape, which is the deepest natural port in the country. The government had earmarked the port of Saldanha Bay for development to encourage exports and, in accordance with race-based social engineering, had designated the region as one for "coloreds." In 1972, an organization called the Cape Midlands Development Association wrote to the government, commending it for designating Saldanha Bay an "economic growth point" that could provide employment for the "growing number of Colored citizens there."[28] But it also protested the fact that the "Bantu" commissioner for the Eastern Province had decided to remove all "bantu" from the Port Elizabeth and Uitenhage areas so that these could be exclusively inhabited by "coloreds." In accordance with the occupational precepts of the color bar, "coloreds" were expected to work as artisans, and the burgeoning motor car manufacturing industry in the Eastern Cape provided ample employment. Protesting against the "synthetic" and forced nature of the arrangement, the Cape Midlands Association stated that "this Association has sound knowledge of the Colored people, and they cannot see them being attracted to the heavy work of steel making."[29] Whether or not the government took these protestations to heart, they provide a sense of the terms of the debate within which racial segregation occurred in tandem with industrial development in Saldanha Bay.

The Sishen–Saldanha railway line ultimately proved to be less impressive than first hoped, and as early as 1971, the major Japanese steel manufacturers downscaled their initial demand forecasts for iron ore. President Nixon's decision to devalue the US dollar in 1971 reduced global demand for Japanese goods, and production at Japanese factories ceased. In addition, Japanese engineers who had traveled to South Africa to assess the development of the port of Saldanha Bay were unhappy with its progress. They estimated that it would be completed later than promised and would cost R300 million more than initially estimated, a cost they believed would be heaped onto the price of iron ore.[30] At the same time, the Japanese government issued blunt condemnations of the apartheid regime, implying that it was unwilling to do business with South Africa.[31] Despite these ominous portents, Iscor's development of its operations at the Saldanha Bay as well as the Sishen–Saldanha railway line proceeded apace because too much had been invested in the project for it to be abandoned.[32]

By 1979, opposition members of Parliament were fully of the opinion that the Sishen–Saldanha railway line was an economic failure. The minister of economic affairs agreed and explained that the railway line had seemed feasible in the 1960s when global demand for South Africa's mineral exports looked set to increase. But the financial downturn brought about by the oil crisis of 1973 inhibited global demand for South Africa's iron ore. The blasted hopes of the Saldanha Bay development threw into doubt the previous certainty with which planners within Iscor and the government had approached their demand forecasts. In a report written a decade later, one of Iscor's engineers, Ben Alberts, who was also involved in developing Iscor's coal mine in the Waterberg, argued that the South African iron ore export market reflected the difficulty of central planning. "The entire forecast process is risky," he wrote, "especially if an attempt is made to forecast a world demand that is subject to political and economic developments in different parts of the world. . . . The history of the iron-ore market bears witness to the influence of over-optimism, uncoordinated planning, the economic situation in the world, and consideration of only some of the factors that will influence the demand for iron and steel."[33]

At the same time that Iscor attempted to improve its ability to export iron ore, it was running out of coking coal for use at its own plants. The best quality coking coal was to be found in the collieries of Natal, which had long been exporters of coal and had supplied the infant colonial economies of East Africa and Southeast Asia in the early twentieth century.[34] Concerned about inefficient mining practices, Iscor had bought over some of these collieries, such as the Durban Navigation Colliery (DNC), which it acquired in 1954.[35] In the 1970s, the mechanization of coal mining grew significantly as coal mine owners increased production to meet a growing global demand for coal. Mechanization also lessened dependence on African labor at a time when the wages paid to African workers were on the rise. This signified a break with the prevailing pattern of the twentieth century, where the abundance of cheap labor gave South African collieries a competitive advantage on the global market.[36] In March 1973, operations at the DNC came to a halt because Xhosas and Pondo workers, who were likely migrant laborers from the Eastern Cape, had allegedly entered into conflict with Basotho workers.

34 ⁓ *Apartheid's Leviathan*

Iscor urged the minister of mines, Piet Koornhof, to intervene so that its coking coal supply was not disrupted. The mine eventually fired "49 Xhosas" and "one Basoto,"[37] but labor unrest continued to disrupt production at various collieries in Natal. In March 1975, Iscor again called on the minister of mines to resolve the situation of persistent rioting at the Hlobane and Northfield coal mines because they supplied 46 percent of Iscor's coal requirements.[38] In response, the minister of mines acted decisively by encouraging the development of more-mechanized, open-cast mining methods, such as longwall mining, in Hlobane to reduce the need for miners to venture underground.

Advances in mining technology during the 1970s also increased the potential for coal extraction. The mechanization of collieries allowed engineers to extract coal they had considered unreachable, thus increasing the size of the country's exploitable coal reserves. Open-cast mining methods improved the extractable coal reserve estimates from twenty-six billion to sixty-one billion tons.[39] In an open-cast mine, coal is scooped from the walls of the coal pit and carried out on the backs of gigantic trucks that crawl along the pit floor. It is less labor-intensive than underground mining but requires greater capital investment in machinery, and this was the method that Iscor opted to use at its mine set amid the Waterberg coalfields, called the Grootegeluk coal mine.

Some of the engineers who worked at Iscor during this time tell the tale of professional struggle and strife out of which emerged technological innovation with profound transformational effects. Michael Deats, a mining engineer by training and an important figure in the exploitation of the Waterberg coalfields, strongly advocated the mechanization of collieries in the 1970s. He came from a family of modest means and secured a loan from a relative for his first year studying engineering at the University of the Witwatersrand in Johannesburg. The diamond company De Beers offered him a bursary for his second year, and the conditions of the bursary set him on the path to qualifying as a mining engineer. After graduating, he worked at De Beers's Kimberly mine for two years before being transferred to its Premier mine near the town of Cullinan.[40] But distraught at the working conditions there—"they treated me very poorly, roughly, cruelly," he said[41]—he left De Beers for Iscor's iron ore mine in Thabazimbi. When he first began work at the Thabazimbi iron ore mine, he was surprised at its thoroughly Afrikaans culture, being himself a native English speaker, but became adept at the language over time and spent seven years as a mining engineer at Thabazimbi.

While still at the Thabazimbi iron ore mine, the mining equipment manufacturer Atlas Copco awarded Deats a bursary to spend two months

The Unlikely Exploitation of the Waterberg ⁓ 35

in Sweden learning about new mining techniques. One of these is known as longwall mining, a more mechanized form of coal extraction than the bord and pillar technique commonly in use at South African collieries at the time. In bord and pillar mining, miners extract as much coal as possible and leave the rest behind in strategic pillars that can support the roof and prevent the surface from caving in. This method is more labor-intensive than longwall mining, since miners work underground, and is considered more inefficient because of the coal that is left behind in the pillars. By contrast, longwall mining employs machines to scoop coal from the coalface, and the roof is allowed to collapse once the coal is extracted. When Deats returned to South Africa, he tried with little success to introduce the technique at Thabazimbi, a failure he attributed to his lack of seniority. He subsequently took up employment at the Iscor-owned DNC in Natal, where he encountered a much warmer reception. The successful mechanization of mining at the DNC made Deats's name within the organization and, by his own account, earned him the reputation as the "doyen of coal mining" in the country. He was later promoted to the position of Iscor's head of mining, an appointment he said was made "with the express purpose of opening up the Waterberg colliery,"[42] and this set him on the path to the Waterberg.

THE SEARCH FOR COKING COAL

In the mid-1970s, Iscor began to fret about its coking coal shortage and urged the government to ringfence certain materials, such as coking coal, for domestic consumption.[43] The increase in global demand for coking coal meant that Iscor suddenly had to pay excessively high prices for coking coal and reduce its overall consumption, which resulted in production losses of "hundreds and thousands of tonnes of steel per year."[44] Iscor had experimented with importing coking coal from the United States but found this too expensive to sustain indefinitely.[45] And the growing wave of anticolonial movements in southern Africa proved inconvenient. Iscor held a 49 percent stake of the Moatize coal mine in the Tete province of Mozambique, but by 1974 the likelihood of Mozambican independence, and the possibility of diplomatic hostility between Mozambique and apartheid South Africa, meant that Iscor could no longer rely on Moatize's coking coal supply.[46] Believing they had exhausted their options, Iscor's board members pinned their hopes on the ownership of coking coal mines and resolved to raise the proportion of coking coal received from its own mines from 15 percent to at least 50 percent by 1980.[47]

Thus began the search for coking coal in the Northern Transvaal, what is today known as the Limpopo province. While Iscor knew that the region

36 ~ *Apartheid's Leviathan*

contained coal reserves, little was known about its scale and quality. Iscor sent prospecting teams to areas known to possess coal reserves, including the Soutpansberg, named after its distinctive salt pans. The Soutpansberg is the northernmost mountain range in South Africa and lies about sixty miles (one hundred kilometers) from the Zimbabwean border.[48] There, Iscor found coking coal of sufficient quality to warrant the development of an underground mine, which came to be called the Tshikondeni coal mine. But the proposed site of the mine lay within the bounds of the former Venda homeland, which worried the Department of Bantu Development because the homeland leadership might claim ownership of the mineral rights if the homeland became independent. The solution was to demarcate the land proposed for the coal mine as a White area, which was nominally outside of the jurisdiction of the homeland authorities, highlighting the actual lack of governmental enthusiasm for the homelands achieving economic prosperity.[49]

While government officials could accommodate a coal mine within the bounds of a homeland, they could not countenance one within the environmental conservation precinct of Kruger National Park (KNP). The proposed site of the Tshikondeni coal mine encompassed a section of the Pafuri camp, which is the northernmost camp of the KNP and the richest in biological diversity. But the government was intractable in its refusal to allow Iscor to exploit the coal in the park, much to Deats's dismay. He described the various cabinet members and officials of the Parks Board as "horrified, absolutely horrified" at the thought of the mine extending underground into the KNP, despite Deats's insistence that the surface would be unaffected. Iscor went ahead with a scaled-down version of the Tshikondeni coal mine, which has since been closed following the exhaustion of the coal reserves. Deats lamented that coal had been left behind within the boundary of the KNP, stating correctly that "the coal doesn't know about the fence."[50]

The development of the Grootegeluk coal mine in the Waterberg was less tenuous than that of the Tshikondeni coal mine, and the area was sparsely populated. Its high temperatures and arid climate prevented easy habitation, and those who settled there did so at great risk, earning it a reputation as a frontier-like region. According to Eugene Marais, a South African essayist who documented his travels through the South African countryside during the early twentieth century, the Waterberg was "associated with all the wonders of unpeopled veld, and to us who were born and grew up on the outskirts of the wilderness it represented the ideal theatre of manly adventure, of great endeavors and the possibility of princely wealth."[51] Its

The Unlikely Exploitation of the Waterberg ⁓ 37

topography reinforced the impression of its social isolation. The Waterberg is bounded at its northwestern edge by the Limpopo River, which also runs along the border between South Africa and Botswana, and at its southern side by the Waterberg Mountain Range. Separated in this way from the rest of the Northern Transvaal, the Waterberg became known as the "zone 'behind the mountain.'"[52]

For White settlers, life in the Waterberg was fraught with sickness and danger. They enjoyed little protection against predatory animals and malaria was rife.[53] The scarcity of water meant that the region was continually afflicted by drought, and with crop farming unsustainable, settlers survived by hunting animals for meat. Importantly, the sparse White settlement in the region meant that the balance of power did not decisively shift from African chiefdoms to White settlers as it did in other parts of the South African countryside during the early twentieth century. But the Waterberg also offered a generous bounty to those who persevered, and hunters partook of the ivory trade, as Marais wrote: "Ivory was then what gold and diamonds became afterwards, and stories were told of bold and lucky hunters killing twenty tuskers in one morning—the value of a principality of land in a few hours."[54]

Geologists had been aware of the existence of coking coal in the Waterberg since at least 1922, after farmers stumbled across coal deposits surprisingly near to the surface while drilling boreholes for water.[55] Prospectors only began to show an interest in the Waterberg after the oil crisis, and when Deats arrived in the Waterberg in the mid-1970s, he found that a few shafts had already been sunk to reach coal seams that were of a suitable quality for the export market.[56] Ultimately, the exploitation of the coalfields required huge financial resources, which a state corporation such as Iscor could marshal, because the infrastructure connecting the Waterberg to the country's various industrial hubs had to be built.

Soon after Iscor began work on the Grootegeluk coal mine, its financial situation deteriorated. By July 1976, Iscor had dramatically reduced its planned capital expenditure and it considered selling some of its assets to gain some liquidity. Its options included selling the Sishen–Saldanha railway line to South African Railways and encouraging the manganese-producing state corporation, Samancor, to purchase the Grootegeluk coal mine, although Samancor declined the offer.[57] Iscor's financial crisis was due to the fact that it had already spent large amounts of money on its expansion program, particularly the Sishen–Saldhana railway line, and 70 percent of its long-term capital expenditure was debt-funded.[58] In its annual report of 1978, the chairman of Iscor described the corporation as heavily indebted,

38 ⁓ *Apartheid's Leviathan*

with its revenue from iron ore exports less than anticipated due to the decrease in Japanese demand for iron ore. In addition, foreign loans became difficult to obtain after the oil crisis and, where available, they came with higher interest rates and tenures reduced from an average of twenty to two years.[59] Persistent political pressure to keep the steel price low meant Iscor could not meet its rising production costs without accruing debt. And since Iscor was then in the midst of its ambitious expansion program, it could not easily curtail its spending.[60]

As Iscor's board members mulled over its financial situation, it became clear in January 1977 that the South African collieries would not be able to meet Iscor's demand for coking coal. They thus resolved to persevere with the development of the Grootegeluk coal mine because canceling the construction contracts would entail the loss of the funds they had already invested while doing little to alleviate the existing coking coal shortage.[61] Determined to proceed against the odds, board members decided to ask the government to contribute R195 million ($11.3 million) toward the Grootegeluk coal mine, a figure based on the cost of a "minimum expansion plan" that utilized what they considered to be "unorthodox" methods, such as short-term loans and lease financing. The government responded that it would provide only R100 million ($5.8 million) and asked that Iscor gather the rest on the loan market.[62] Iscor eventually managed to raise a further R295 million ($17.1 million), stating that this was achieved "once again at the risk of funding Iscor by methods as would not, in all instances, reflect sound financial policies and principles."[63]

Iscor was entering unchartered financing territory, one it lacked the institutional tools to navigate. At board meetings, its officials constantly discussed more creative fundraising possibilities from various, dispersed entities, including the government, fellow state corporations, private financiers, and construction companies. One of the financing options, which proved too expensive in the end, involved reaching "reasonable terms" with an obscure financial entity called Southern Life Association to raise R100 million ($5.8 million).[64] But "the precedent setting nature of this type of financing" proved too radical for its time because it required prohibitively expensive computer-assisted calculating to manage "the vast amounts involved." One of the board members suggested that Iscor employ a special consultant with sophisticated analytical expertise to improve the corporation's capacity to manage new sources of financing.[65]

In July 1977, Iscor arrived at a new scheme: to sell the "non-metallurgical" coal, or the bituminous coal, of the Grootegeluk coal mine to its sister state corporation Eskom, thereby indirectly sharing the costs of operating the

The Unlikely Exploitation of the Waterberg ⟿ 39

mine.[66] This meant that Eskom would have to build a power station in close proximity to Grootegeluk. But Eskom was initially extremely reluctant to enter the Waterberg, and Deats described them as about "as interested as my cat is in running after dogs."[67] So Deats cajoled various cabinet ministers during the John Vorster regime and took them to the Waterberg to demonstrate the importance of building a power station there.[68] Eventually, the government and Eskom overcame their initial indifference and agreed to construct the power station in the Waterberg, for reasons discussed in greater detail in chapter 3.[69]

For one of the founding managers of the Grootegeluk coal mine, Joe Meyer, the steely determination that drove Iscor's engineers to the Waterberg held within it a nation-building imperative: "We had huge challenges at Thabazimbi [iron ore mine], and that made us persevere, *vasbyt,* in a way that you can't believe so that the country could grow. . . . I mean in those days the people came from the farms after the Rinderpest [cattle-killing disease], it was just after the Second World War, people just wanted to improve their conditions again, to stabilise after a very challenging time where a lot of people lost everything they owned."[70] Meyer recalled his childhood in a rural area of the country, where he would have to walk long distances barefoot each day between the primary school he attended on a farm and his boarding house. After secondary school, he enrolled at the University of Pretoria to study civil engineering in the evenings. Lacking funds to complete his degree, he approached Iscor for a bursary to fund his last two years of study. Iscor, however, was only concerned with funding mining, electrical, mechanical, and chemical engineers to support the mining and steel manufacturing sectors. Meyer recalled waiting in an intimidating waiting room, where he watched applicants ahead of him in the queue leave empty-handed. During his own interview he insisted on becoming a civil engineer and snidely remarked that Iscor would find this a valuable expertise considering the dire state of their infrastructure.

Against the background of cost cutting within Iscor as a whole, Iscor's engineers felt the pressure of parsimony at the Grootegeluk coal mine. Second-hand equipment was used where possible, and in 1978, Deats gained the board's approval to import second-hand "stackers" from halfway across the world. This equipment had first been used for the construction of the largest earth-filled dam in the world, the Tarbela Dam in Pakistan.[71] Meyer describes Iscor's activities in the town during the 1970s as severely constrained by the lack of funds. But Meyer was still expected to produce the necessary infrastructure for the mine and in the town. This included the development of the Mogol Club, a recreational hub with various sports facilities, intended

40 ～ *Apartheid's Leviathan*

to serve the staff and the surrounding White community in the otherwise derelict portion of the Bushveld.[72]

In keeping with Scott's view of authoritarian high modernism in the twentieth century, the apartheid government relied on the know-how of scientists and engineers to realize their vision of modernization and control. This vision was racially delimited and consisted of improving the lot of poor Whites, who would enjoy the benefits of modernization, and controlling the mobility of Africans in nominally White South Africa. But this vision of modernization could not be realized by the strength of the state bureaucracy alone. Since industrialization required sophisticated scientific expertise, the project of national industrialization was driven by the state corporations. It appeared at first to perfectly fit the imperatives of the apartheid racial order. The conditions of financial prosperity in the 1960s, at least in comparison to the drying up of funds after 1973, led to supreme confidence in Iscor's planners, who predicted an increase in global demand for its iron ore and in the domestic demand for its steel. Where Iscor established new steel manufacturing plants, which offered the potential to create new industrial hubs in undeveloped regions, the government ensured that these were structured to further the aims of the apartheid racial order. For example, Iscor's steel plant at Newcastle served the purposes of industrial decentralization, an apartheid policy that conferred an economic purpose on the homelands as pools of surplus labor. Iscor's development of Saldanha Bay was similarly in accordance with the government's vision of the racial job reservation best suited for the region, which saw the area being emptied out of Africans. But while there were times of concord and agreement between the state corporations and the government, there were also times of discord, as in the government's refusal to allow Iscor to situate a coal mine within the bounds of the KNP. In this way the relationship between Iscor and the apartheid government can be characterized as both autonomous and immersive.

As planners within the government and the state corporation expanded the country's industrial facilities, the oil crisis of 1973 came as a major shock to the system. Iscor struggled to access the same loans from overseas creditors it could before the oil crisis—loans that had been more conducive to its long-term infrastructure development—and the government did not readily fill the funding gap. Iscor's new situation of funding scarcity threatened the viability of the projects it had already begun, such as the Grootegeluk coal mine in the Waterberg. But since the cost of turning back was higher than tenaciously proceeding, Iscor resolved to search for "creative" sources

The Unlikely Exploitation of the Waterberg ⁓ 41

of funds. The state corporations' involvement in the Waterberg was hardly a strategic imperative of the government, and government officials reluctantly assisted where required. The solution, ultimately, came from Eskom agreeing to locate one of its new power stations in the Waterberg.

The oil crisis of 1973 is seen to have initiated neoliberalism across the African continent, in South Africa at least, but it did not immediately result in the end of the state-led development project. While the scarcity of funds placed Iscor's expansion plan on shaky grounds, it initiated a new configuration of cooperation, setting the stage for the exploitation of the region's coal reserves in the era of democracy. Whether unwittingly or not, Iscor fitted into the Waterberg's historical mystique as a frontier region and unexplored treasure trove that would only yield to the staunch-hearted. Few at the time would have foreseen that by the end of the twentieth century, the Waterberg would become the focus point of a multibillion Rand infrastructure project.

2 ~ The Taming of the Waterberg

ISCOR'S ENGINEERS centered their activities in the Waterberg on the town of Ellisras, which lay close to the Waterberg coalfields and was so small it was barely distinguishable from the surrounding bushveld. While lacking appropriate municipal infrastructure and diminutive in size, the town had been riven with conflict among farm owners since the 1960s about the direction that urban development should take. And, in contrast to the pattern elsewhere in the Northern Transvaal, by the mid-1970s Africans had not been forcibly removed from White-owned land to the nearest homeland. The mostly diffuse local authority was in turn a product of the sparsity of the White population and the absence of a thriving commercial center around which a town could form.

During the nineteenth century, the sparse White settlement in the Waterberg, due to its hostile environmental conditions, meant that the government was not called upon to uphold White dominance of the African chiefdoms in the region. Elsewhere in the countryside, by the beginning of the twentieth century, the White settlers who had made their way from the Cape into the South African interior had mostly subdued the once powerful African chiefdoms they encountered. The 1913 Land Act cemented African land dispossession, and legal restrictions on African ownership of land only intensified in ensuing decades, as did the policing of African mobility. The authority of White settlers on the ground—and their right to land in particular—was bound to the juridical power of the state, first under the suzerainty of the Zuid-Afrikaanse Republiek and then under the state structure put in place by Lord Alfred Milner in the aftermath of the South African War (1899–1902).[1]

The apartheid regime entrenched racial segregation in government policy, and the Group Areas Act of 1952 gave life to the Group Areas Board, which oversaw segregation in big cities and small towns across South Africa. The Group Areas Board set the mold of urban segregation, within which only Whites enjoyed freedom of movement in central business districts, while Africans resided in adjacent townships or in distant African homelands. But the frontier-like region of the Waterberg had yet to encounter the full power of the segregationist state when Iscor's engineers arrived in the mid-1970s. Having long been a scarcely populated region, the Waterberg had escaped the government's regulatory eye for much of the twentieth century.

Iscor's arrival in the Waterberg coincided with a period of introspection within the National Party (NP). A factional battle that had long been brewing within the NP[2] boiled over in the aftermath of the Soweto uprising of 1976 as the apartheid government adopted what Deborah Posel has termed a new "language of legitimation."[3] The verligtes embraced commercial imperatives, even if this meant relaxing racial segregation, much to the chagrin of their *verkrampte* counterparts, who derided the embrace of individual enrichment and saw it as a betrayal of the ideal of a classless Afrikaner unity. The farmers of the Northern Transvaal, of which the Waterberg was a part, were an important electoral constituency, and the NP held campaign rallies across the region in an attempt to stem party defections. In one such rally, held in Ellisras in November 1978, Prime Minister P. W. Botha appealed for party unity in the face of mounting claims from his own verligte camp that a split might be an essential component of the party's "purification" process, and he defended the need for policy changes.[4] The rift eventually grew serious enough to cause a breakaway led by Andries Treurnicht, who had served as the NP's parliamentary representative for the Waterberg since 1971. Treurnicht formed the Conservative Party in 1982, and the party opposed the concessions that the NP had begun to make to big capital as well as its apparent relaxation of racial segregation. As apartheid changed its form in the 1980s, the Group Areas Act remained as the last pillar of grand apartheid before it was finally repealed in 1988.

In Ellisras, despite its apparent verkrampte hue, the Group Areas Board had not implemented racial segregation when Iscor first arrived. Africans continued to live on White-owned farms far longer than was the norm in other parts of the Transvaal. Local government officials believed that the forced removals of Africans to the Lebowa homeland would cause a labor shortage and so harm the existing economic fabric.

Within the government's conception of regional development, the homelands served as pools of African labor. A 1965 memorandum written by an unnamed government official about the plan to develop the Ellisras region stated that it was government policy to encourage autonomous regional development in line with the then-normative pattern in "Western democracies." In this pretension of minimal government control, autonomous regional development was a systemic effort to ensure the "optimum development" of the region's natural resources, including labor. The people and things within the boundaries of the region would be reduced to units within their appropriate categories, and the interaction of these "resources" would ensure development. Agents at the local level were to make their views known through representative organizations, and the government weighed in through various interest-bearing ministries.[5] Since Africans lacked any rights of citizenship in the White towns that contained the infrastructural bearers of modernity, they fit into the calculus of the regional planners only in their role as laborers to be shepherded to the homelands, where they were officially resident.

Forced removals can be seen as an important part of the praxis of authoritarian high modernism, and they fit into the apartheid government's systemic vision of regional development. Soon after Iscor arrived in Ellisras, African families living in the vicinity of the town were forcibly removed to an African township in the nearest homeland of Lebowa. But Iscor, armed with its technological expertise, did not initiate the removals. Rather, Iscor's arrival heralded population growth in Ellisras and the government was forced to turn a regulatory eye to the town to ensure its orderly development—in line with the "modern" city in the apartheid imaginary. Apart from containing the appropriate infrastructure, the modern city was one that was racially segregated. This, in turn, depended on an underlying economic prosperity so that farms and businesses could survive despite forced removals, which, after all, moved African workers further away from their places of employment. In Ellisras, forced removals were the product of negotiation and compromise between the Group Areas Board, local and provincial authorities, White residents of Ellisras, and Iscor. While Iscor provided the infrastructural tools for the creation of Ellisras as a town worthy of government regulation, it was chiefly concerned with the operations of its coal mine, and

The Taming of the Waterberg ⁓ 45

there is little to suggest it was deeply invested in adhering to the dictates of the Group Areas Board. In the end, Iscor's activities enabled the extension of governmental power from the capital to the border region of Ellisras in a mediated and complex manner.

EARLY CAPITALIST DEVELOPMENT

In the early part of the twentieth century, when White settlers first populated the region, the Waterberg's heat and aridity made crop agriculture difficult, and residents relied instead on hunting wild animals for subsistence. Water could only be found in the slender and temperamental flow of the Mogol or Mokolo River, which is a tributary of the Limpopo River, and the most desirable land was on the riverbank. The town was founded, as the story goes, by two settlers, Patric Ellis and Jan Erasmus, who merged parts of their last names to form the name Ellisras. Ellis grew up in the Marico district, about 185 miles (300 kilometers) southwest of Ellisras, and was a member of the landless stratum of Afrikaner settlers called *bywoners,* who worked the land as tenant laborers. He migrated to the Waterberg in search of independence from the sort of overlordship that prevailed in the Transvaal and settled on the banks of the Mogol River on a farm that came to be called the Waterkloof farm.[6]

In the 1930s, South African Railways and Harbours placed a bus stop on the Waterkloof farm, for a route connected to a railway line that passed through the town of Vaalwater, which lay to the south of Ellisras and ran northwards to the town of Stockpoort, which is the last South African town before the border with Botswana. This transportation network placed the incipient town on the map. The railway line also enabled the passage of merchants selling household wares to rural districts such as Ellisras, and the railway stop formed the basis of a rudimentary economic center.[7] Barter was the prevalent form of exchange in the region. White settlers hunted game in the bushveld and exchanged dried meat for agricultural crops, such as grain and peanuts, with the nearby African chiefdoms. A single storekeeper served as the only contact with cash in the district, and Ponk Ellis, the grandson of Patric Ellis, described White settlers as having exchanged chicken, bones, and "anything you could give him"[8] for household supplies. In this way there existed in the early twentieth century a residual interaction on terms of relative parity between Africans and Whites in the Ellisras region, departing from the general pattern of conquest and control that had already occurred across much of the Transvaal by the 1930s.

46 ～ *Apartheid's Leviathan*

But Ellisras soon caught up with developments in other parts of the Transvaal. For Africans in the district, the twentieth century saw the gradual erosion of their economic autonomy. Sam Sekati, a local businessperson whose parents were born in Ellisras, described a historical process of land dispossession through bureaucratization.[9] White settlers demarcated the plots of land by pegging their corners and then drawing up title deeds to certify their ownership of the land. Backed by the juridical authority of the state, they gradually acquired cattle from members of the African chiefdoms, who had historically used the entire region to graze their cattle, unconstrained by the boundaries of titular plots. The mobility of cattle across grazing pastures helped to prevent overgrazing and soil erosion because it allowed the grass time to regrow. Sam Sekati described the process of dispossession as a gradual one, where White farmers settled on the land in apparent parity with their African neighbors for a time before claiming ownership of the land. Then, in positions of authority, they would order individual Africans to reduce their cattle stock. This is a familiar tale in the region, with one interviewee lamenting the fact that, while his father had once owned a large herd of cattle, he was eventually left without even a chicken to his name.[10] As Sekati related:

> You must remember, when the white people arrived here, they didn't have anything. And then you will find one family has got hundreds or more cattle. And then the white guy will come, he will stay with you, with your neighbor, and after a certain time he will tell you, "I've got the farm." And then if you agree to stay on that farm, maybe for a year, then next year he will tell you, "I think your cows are too many, you must reduce your cattle." And then he will buy your cattle for far less—he will keep on reducing your cattle until you are left with nothing. Maybe you are left with five, and then he'll say, "That's enough." So that's what happened. . . . And then you will go to another place. You have one hundred, two hundred cattle, and then you are left with only four or five. And their tactic was that if you refuse to sell your cattle and then you decide to move to another place, another farm, the farmers will communicate. They will say, "If a black person is coming to your farm, don't allow him to bring so many cattle." So that was the strategy, that there is nowhere to run.[11]

But this wealth disparity did not immediately translate into racial segregation, at least not in the sense mandated by the Group Areas Board.

The Taming of the Waterberg ⁓ 47

Willie Loots, who grew up in Ellisras, recalled the presence of informal African settlements on land alongside the road that led to Polokwane before the commencement of forced removals.[12] Along with other scattered pieces of marginal and under-capitalized farmlands in the Transvaal, which the government called "black spots," White farmlands in Ellisras continued to see African occupation as late as the 1970s. In 1968 government officials complained about a group of "unlawful" squatters that resided in close vicinity to the incipient town, some of whom had lived and worked in Ellisras for eighteen years. The group included Hereros from South West Africa (Namibia) who had immigrated in 1918 along with other migrants from unnamed regions, and who were described only as "dorsland trekkers" or "desert migrants." They could not easily be removed because of their lack of ethnic homogeneity, which complicated their incorporation into the nearby Lebowa homeland, nominally meant for Northern Sothos.[13] In addition, the government officials believed that forced removals would have resulted in a labor shortage because daily transport from the Lebowa homeland was prohibitively expensive.[14]

DROUGHT AND AGRICULTURE

The farmers of Ellisras in the twentieth century were wholly dependent on the fickle flow of the Mogol River, which could run dry for years at a time. During these dry years, farmers would suck water from its deep sand bed, and that sufficed for household use.[15] The bushveld of the Northern Transvaal is generally arid, consisting of hardy, thorny foliage, and cattle ranching was generally a more profitable occupation than crop agriculture. The White farmers of the Transvaal and the Orange Free State had carried the NP to a surprising victory at the polls in 1948, and once in power, the NP prioritized improving their prospects. Their intervention focused specifically on small-scale White farmers, continuing the earlier governmental tradition of eradicating poverty among Whites. For instance, the apartheid government implemented a number of interventions to marketize the agricultural production of small-scale farmers. This included the development of agricultural cooperative societies across the countryside, regulated by marketing boards, which centralized the marketing of a particular crop in any given district and ensured that all farmers were paid in two installments—the first after they delivered crops to the cooperative's facilities and the second after the crops had been sold.[16] The government also built supportive infrastructure, including huge concrete grain silos and dams. Dam building was particularly extensive; William Beinart estimates that by 1970, South Africa's

48 ~ *Apartheid's Leviathan*

proportion of "channeled water" for irrigation was among the highest in the world.[17] Agricultural cooperatives prospered in the town of Potgietersrus (today known as Mokopane), which lay about 125 miles (200 kilometers) southeast of Ellisras, and supported extensive tobacco farming in the region.[18] The apartheid government also extended its assistance to farmers in the Ellisras region, and Ponk Ellis recalled the government assisted farmers by drilling boreholes on their land and parceling out plots of land to farmers to encourage White settlement.

But these governmental efforts could not mitigate the effects of a major drought that occurred in the 1960s in the Northern Transvaal. A government inspector who visited Ellisras in 1965 found that a drought that spanned 1961 to 1964 had decimated farmers' livelihoods.[19] Only those farmers who possessed cattle were able to survive the drought financially, though their cattle stocks had reduced and their debts had grown. To make matters worse, foot-and-mouth disease spread among cattle immediately after the drought, and farmers were unable to sell any of their sick cattle to repay their loans. Where possible, farmers quickly sold as much as half of their cattle for liquidity and to prevent extensive soil erosion.[20] In the end, financial assistance from the government proved a lifeline for struggling farmers. Johan Pistorius, one of the members of the local committee tasked with channeling government aid to struggling farmers, traversed the Ellisras district to investigate the state of the surrounding farmlands. He describes the 1970s as a period of great strife for many: "We visited a lot of people. We met a guy—he will tell you with tears in his eyes, 'I'm here for twenty or thirty years and I never made a loss, every year there was a profit. But the profit couldn't handle the inflation.' So costs went up and they couldn't survive."[21] Government aid took the form of agricultural credit that saw farmers receiving interest-free loans for five years followed by a fixed interest rate of 2 percent. Pistorius's committee determined who qualified for the loans and assessed the merits of individual farmers on a case-by-case basis and enjoyed some success in keeping farmers on the land.

Cattle ranching became increasingly unprofitable over time because of overgrazing and extensive soil erosion. In addition, cattle farmers struggled with a persistent plant that was poisonous to cattle, known by its Afrikaans name of *gifblaar*.[22] Farmers found game farming to be a more profitable use of the land, beginning a process that would eventually see the Waterberg becoming home to internationally renowned game farms. The transition from cattle to game farming is a characteristic feature of the changing economy of the Northern Transvaal during this period.[23] Game farming did not require the large plots of land that cattle ranching did because it depended on very few, but expensive, animals.[24] In 1974 Pistorius was among the first farm owners to erect a game fence around his property. The simple act of fencing the property ensured that the animals within it belonged to the property owner and so allowed game to be monetized. Fortunately for the game farmers who invested in specialized and rare species, there was a strong international demand for hunting southern African game. In the 1980s, game farm owners began to advertise globally, particularly in the United States, to attract foreign hunters to their game farms to hunt more exotic, expensive game. To this day, game farming is a thriving tourist magnet in the Waterberg and has attracted foreign investment from as far afield as Saudi Arabia.

TOWN PLANNING

When Michael Deats, an engineer at Iscor responsible for the corporation's mining operations in the Waterberg, arrived in Ellisras in the 1970s, he found a town that was signaled only by an unremarkable smattering of buildings. "There was a bottle store," he said. "And there was a little kind of hotel—and when I say little, I mean little. And a few houses. And I think there was a magistrate." The town of Ellisras had developed in a halting fashion before Iscor arrived, and its residents paid little attention to the precepts of town planning. John Oswald Whelpton, a farm owner, first applied to the government to establish a town on his plot of land, arguing that the main towns of the Northern Transvaal were too far away to effectively service the White farmers of Ellisras. Government officials, who visited Ellisras in subsequent years to consider the application, bemoaned the directionless development that had already occurred. In June 1959, the commissioner for urban areas made a trip to Ellisras, and on arrival at the few desolate buildings that together constituted the center of town, he found there was no one to greet him.[25] After some inquiries, the magistrate of the town of Nylstroom, which lay ninety-three miles (150 kilometers) from Ellisras, alerted a representative from the Ellisras Farmers' Association to his presence, and the man finally appeared twelve hours later.

The commissioner's annoyance at the lack of a reception seemed to color his report, and he criticized the "toadstool development" of the few buildings in the town, which had sprung up in complete disregard for urban zoning regulations, resulting in irreversible errors such as a church where a factory should have been and houses in commercial districts. Importantly, the government insisted that the infrastructure for the provision of basic

50 ～ *Apartheid's Leviathan*

services should be in place before it could officially recognize the new town. Regulations included the presence of a reliable water supply for household requirements as well as for firefighting purposes. Borehole water would not suffice, because the plots of land were too small to ensure that septic tanks for sewage treatment were placed a suitable distance from the boreholes, thus increasing the risk of underground water pollution. Whelpton had not made provision for dumping grounds, a cemetery, or a working sewage system in his initial application.[26] In addition, the Department of "Native" Affairs[27] insisted that any emergent town had to demarcate an area to establish an African location.

Whelpton seems to have satisfied the water provisions requirements at least, and in 1960 the government declared Ellisras to be a township, situated on the farm Waterkloof. But Whelpton remained consistently at odds with the government, and in 1964 he launched legal action against the government's restrictive urban zoning regulations which, he argued, meant that businesses could not be situated where he wished.[28] Whelpton left Ellisras soon thereafter, no doubt in frustration, to settle across the border in Botswana. He sold his farm to a private company called the Joubo Development Corporation (JDC), which assumed responsibility for the provision of basic services, particularly water, to the scattered townsfolk. But residents of the town complained of shoddy service and irregular water provision in 1965.[29] After investigating the matter, the head of the Department of Local Areas reported that the JDC had been charging excessively high rates for cloudy water.[30] During the rainy season in the summer of that year, the water pump had been submerged in a flood from the Mogol River, but it was defective even under normal working conditions. In 1971 a different set of residents applied to have a town developed on their property, resulting in the formation of "Ellisras Extension 2," a township that lacked a road link to Whelpton's already existing township.

Iscor and Eskom put an end to these disparate, uncoordinated efforts to develop a town by offering to provide the infrastructure for water and electricity provision. In 1976, Mike Deats, acting on behalf of Iscor, promised to provide the town with a portion of the water and electricity that Iscor intended to channel to the Grootegeluk coal mine from the Hans Strijdom Dam, which was then still under construction, and to supply the town of Ellisras with a part of the electricity it would receive from Eskom.[31] Assured by this guarantee from Iscor, the Peri-Urban Areas Board, a provincial governance structure for small towns that had retained a loose oversight over Ellisras since 1965, rejected the numerous applications

The Taming of the Waterberg ⁓ 51

it had received from residents of the district to develop a town on their properties.[32]

The fact that the Peri-Urban Areas Board had delegated responsibility for water and electricity to Iscor caused some consternation among already anxious applicants. One of these applicants was the JDC, which had formed a consortium with a company called the Transmogol Beleggings Beperk (Transmogol Investment Company). They were anxious to establish the town on an extension of the original Ellisras town, promising to source water themselves from the Mogol River, which they felt could be sufficient to service the proposed town with enough water.[33] But the board refused to approve their application, insisting they first demonstrate that the necessary utilities were available.[34]

When it was eventually operational, the Hans Strijdom Dam was the first reliable and consistent irrigation source for farmers in the region.[35] While owned by the state, Iscor maintained the dam and controlled the flow of water to the Grootegeluk coal mine, the town of Ellisras, and to farmers on a quota basis. In this way, Iscor and Eskom overpowered private efforts to provide water and electricity to the town and cemented the dependence of the town's administration on their infrastructure for municipal service provision.[36] Iscor provided the appropriate infrastructure to insert the town into the orbit of the government-sanctioned vision of modernity.

ISCOR AND ONVERWACHT

When Iscor first arrived in Ellisras, it struggled to acquire land. Joe Meyer, who was intimately involved in the day-to-day operations of the coal mine, described residents as deeply suspicious of the miners who threatened to disrupt the rhythms of rural life.[37] Iscor's acquisition of land was further complicated by the fact that the existing plots of land ran for kilometers in narrow strips from the banks of the river.[38] Through processes of settlement and sub-division over the course of the twentieth century, the Waterkloof farm had fragmented into many small but long plots of land.[39] Nonetheless, Iscor eventually found a willing seller in the owner of a massive farm called Onverwacht, which lay about three miles (five kilometers) north of the Waterkloof farm and a greater distance from the banks of the river.

Iscor used the Onverwacht land to build the Grootegeluk coal mine, the African township, residential housing for its own staff, and a business district.[40] In so doing, Iscor split Ellisras's commercial life in half—resulting in the creation of two separate and antagonistic chambers of commerce,

52 ⁓ *Apartheid's Leviathan*

each representing different central business districts.[41] Some residents remained disgruntled over Iscor's intrusive presence, and in the 1980s they complained about Iscor's proposal to build a road that connected the two urban commercial centers.[42] One resident complained that building the proposed road would fragment his property and destroy valuable installations, which included a sprinkler irrigation system that would fall beneath the proposed tarred road. He proclaimed that residents were unimpressed by the economic boom that Iscor promised to bring and preferred to be left alone with their own problems. Rejecting the idea of a shared regional prosperity, he wrote that Iscor should develop where it liked and leave the old town and property owners to their own devices, even if this meant that it became a ghost town.

FORCED REMOVALS AND HOMELAND CONSOLIDATION

The transformation of Ellisras into a suitably modern town was tied to public health concerns. The first permanent official tasked solely with the town's administration in 1966 held a broad public health portfolio that included administering smallpox and polio vaccinations and overseeing measures to control the spread of tuberculosis, malaria, and bilharzia. Apart from ensuring that the Department of Health's requirements were met, he also functioned as a building inspector and a "Bantu" Affairs[43] official, and he had to maintain the sanitation infrastructure, such as drainage and rubbish collection.[44]

With the Waterberg coalfields firmly in sight of Iscor's engineers, the town swam into the view of the Group Areas Board. An inspector from the board who visited the town in 1966 found to his dismay that African families lived in dispersed settlements across the district, occupying farmlands with the consent of the owners in tenancy arrangements. He insisted that an African township be established and met with the Ellisras Local Area Committee to set aside land. Despite some consensus, the committee remained irresolute on the matter, much to the inspector's annoyance.[45] But local officials could only drag their feet for so long. By the mid-1970s, Iscor's arrival was no longer in question, and the need for "ordered" urban development in response to the coming industrialization became more urgent. Officials cast a renewed eye on the African families scattered on the land in the general vicinity of the town. In 1975 the commissioner for Bantu Affairs reported that one farm owner was in the process of evicting African families from his land, and he feared that further evictions would hasten a desperate labor shortage for the town. Chastising the Peri-Urban Areas health inspector, the

The Taming of the Waterberg ～ 53

commissioner reiterated the need for the speedy development of an African township to allow the removals to occur.

There is a clear link between the industrialization thought to follow from Iscor's activities and the intense concern over public health among government officials. In March 1976, the health inspector reported that generally unhygienic conditions prevailed in the informal settlement and regularly generated complaints from some White residents of Ellisras. The "squatter camp" in question was an informal settlement on the outskirts of the town known as the Pahama location. The imminent arrival of Iscor heralded a more chaotic residential situation that would create the perfect conditions for the spread of infectious disease such as polio, chicken pox, tuberculosis, and diphtheria. The health inspector also visited two farms upon which African families resided and reported that, while the immediate surroundings of the "Bantu" huts were clean, the "Bantus" on one of the farms dumped their rubbish over the border fence. In addition, African residents drew their drinking water from the Mogol River and suffered from a general lack of sanitary services. He hoped the commissioner of Bantu Affairs would cooperate in providing latrines and proper water supply. But more importantly, as the final solution for the sanitary problems associated with informal dwellings, the health inspector urged the speedy construction of the African township.[46]

While health officials continually expressed their concern, the commissioner of Bantu Affairs for the Northern Transvaal held the most sway in expediting the forced removals. This was in line with a broader national move to eradicate "black spots" and consolidate the authority of the homelands. He was particularly concerned that his office had no control over the Africans living on White-owned land and thus had no authority to dictate how the White farm owners treated them. At that time, the Bantu Trust was about to purchase land in the vicinity of the town of Marken, roughly forty-five miles (70 kilometers) away from Ellisras, for the construction of an African township. This proposed site was situated inside the Lebowa homeland and would eventually become known as Steilloop.[47]

In 1976, Iscor promised to construct a hostel for single males in Ellisras, with an allowance for 3 percent of African employees to reside as families. In the meantime, the Pahama settlement was declared a temporary location and its residents earmarked for later relocation to the homeland.[48] Homeland consolidation held a financial rationale for the apartheid government. Rising inflation in the 1970s encouraged the government to lower its direct spending in the homelands and increase the tax revenue of the homeland

54 ～ *Apartheid's Leviathan*

authorities. This would work only if African workers paid taxes to their respective homeland governments. African workers who resided in White areas were of particular concern because it was not possible to collect taxes from them.[49]

Homeland consolidation also coincided with the militarization of South Africa's borders, which was being prioritized in response to the independence of southern African countries. In Prime Minister P. W. Botha's vision of a homeland "constellation," the African leaders of nominally independent homelands would be dependent on, and thus friendly to, the apartheid government. Since the homelands were situated near South Africa's borders, they could collectively function as a buffer zone in the event of an invasion from the armies of African countries to the north.[50] Botha also hoped to dilute African nationalist aspirations within South Africa through encouraging homeland independence. The reckless ambition of this scheme cannot be overstated, and Saul Dubow writes of the homeland system that "influx control, population resettlement, and Bantustan creation involved a heady mix of self-deluding ideological vision and cynical calculation."[51] Some homeland leaders accepted their newfound independence with great aplomb, such as Kaizer Matanzima of the Transkei and Lucas Mangope of Bophuthatswana. By contrast, successive leaders of the Lebowa homeland adopted a consistently hostile stance to the prospect of independence.

This grand scheme of homeland defense did not obviate the need to defend the borders against guerrilla incursions from neighboring African countries. Before Iscor arrived in Ellisras, the region already held a military airport housing fighter planes and other defense aircraft. Iscor and Eskom's activities assisted in bolstering the borders of White South Africa and preventing the rapid depopulation of the countryside. Following Mozambican independence and the coming of African majority rule in Zimbabwe, the northeastern Transvaal border areas, including Ellisras, were considered White "outposts" by 1982.[52] In 1978, a body called the Civil Defense Unit had reportedly ensured that farmers could access medical supplies, food, fuel, and ammunition in a secret location in case of an invasion. The District Agricultural Union applied to the minister of agriculture for government loans at reduced interest rates so that the remaining farmers could purchase two-way radios and erect security fences around their farms.[53] Many White farmers had left the district entirely. A newspaper article in *Die Transvaler,* quoting figures from the District Agricultural Union, showed that 60 percent of the farms in the Ellisras district had no White farmers resident on the property. In the nearby town of Thabazimbi, the figure was higher, at

The Taming of the Waterberg ⁓ 55

75 percent. In addition, nearly four thousand square miles (ten thousand square kilometers) of farms in Ellisras had been sold or abandoned by White farmers, resulting in the much dreaded depopulation of the countryside.[54]

Newspaper articles from the 1980s detail incidents of bombings in Ellisras and the surrounding farmlands. In one case, police shot and killed a suspected "terrorist" after he allegedly attempted to throw a hand grenade at them.[55] In 1986, a land mine at a farm in the Stockpoort district exploded and killed an elderly couple. The chairman of the Ellisras Farmers' Association told the *Star* that unoccupied farms in the district provided a safe haven for "illegal immigrants" and were breeding grounds for terrorist activity.[56] In 1984 the Ministry of Constitutional Development and Planning approved R34-million ($1.8 million) in spending over two years to "stabilise" the border region of the northwestern Transvaal by improving its "agro-economic conditions." These improvements included infrastructural development such as the electrification of the district and road building schemes.[57] A newspaper article describing farms on the Botswana–South Africa border as being on the "frontier" stated that it was "difficult to find two white-occupied farms in a row. One white ranch manager told [the journalist] he did not have a white neighbor for 20 kilometers [12 miles]." Nonetheless, seventy-year-old Piet Pretorius, who was interviewed for the article, denied feeling under any threat, downplaying the paranoia of the "townspeople" and describing his friendly relationship with the chief across the river.[58]

The government soon commenced with a wave of "black spot" removals, which, in the case of the Lebowa homeland, elicited protests from its leadership as well as from the affected communities. In a speech delivered in 1987, the newly elected chief minister of Lebowa, Mogoboya Nelson Ramodike, commended the leadership of his predecessor Cedric Phatudi, describing him as a liberation hero: "Under his courageous comment, Lebowa people rejected so-called independence which he appropriately described as self-strangulation."[59] He called for the legislative assembly of the homeland to reject the notion of independence as part of the constitution of the Lebowa homeland. Lebowa was not an ethnically exclusive homeland, he argued, and would house people of different tongues within its borders. Despite the dissent, Lebowa remained the demarcated homeland for Northern Sothos, even though the people of the region did not necessarily define themselves as such. In 1985 members of the Seleka community, which was near Ellisras, launched legal action against the minister of justice of the Lebowa homeland. They protested his declaration in 1969 that their so-called "tribal" land (called Beauty) would be a part of Lebowa. In addition, the Seleka resented

56 ～ *Apartheid's Leviathan*

their official incorporation into a Northern Sotho homeland because they argued that, as Tswana speakers, they did not fit ethnically.

Eventually, the African families living in informal settlements on White-owned farmlands surrounding the town of Ellisras were banished to the Lebowa homeland. The removal was a traumatic one for many. Sam Sekati, whose family was removed to the homeland during this period, had difficulty describing the experience, saying only that it was "bad, very bad." In 1979, the *Sunday Express* detailed the removals of an estimated twenty thousand Africans across the Northern Transvaal, including those who resided in African locations in the proximity of the towns of Naboomspruit, Nylstroom, Vaalwater, Louis Trichardt, and Ellisras. When these reporters visited Ellisras to report on the forced removals there, they initially struggled to locate the site of the informal location: "We had difficulty in finding the site of the former Ellisras location, two kilometres [one mile] outside the town. At last we realised we had driven past it—it had been bulldozed flat, the ground surrounded by a two metre [six-foot] game fence. A few ostriches were stalking among the ruins."[60] The newly erected hostel, painted a "vivid pink," was situated twelve miles (twenty kilometers) out of town, close to the Grootegeluk coal mine and already populated only with African men. While the government subsidized the cost of houses in the homeland, according to Sekati, living conditions remained unbearable. As a result, many who were financially able departed as soon as the dissolution of the apartheid regime rendered racial segregation in residential areas obsolete.

Iscor arrived in the Ellisras district in the aftermath of a major drought that had destroyed the scanty livelihoods of many White farmers. The region had escaped the regulatory eye of the government for much of the twentieth century, and residents who attempted to form a small town center struggled to develop the infrastructure necessary to deliver services to residents of the town in accordance with government prescriptions. In addition to the infrastructural requirements, a modern town in the eyes of the government was a racially segregated one. Enshrined in the activities of the Group Areas Board, spatial segregation required the forced removals of disenfranchised families and communities from racially mixed areas to racially demarcated regions. This huge feat of social engineering is akin to the characteristics of authoritarian high modernism that Scott has described.[61]

While forced removals of African people from informal settlements coincided with Iscor's arrival in the town, it is important to note that Iscor did

The Taming of the Waterberg 57

not necessarily initiate the forced removals. Rather, Iscor's activities signified industrialization and the prospect of greater population density in Ellisras, rendering racial segregation an urgent necessity. Iscor's infrastructural network overpowered residents' disparate and fraught efforts to control the direction of the town's development. At the same time, the apartheid government grew more concerned about protecting the country's northern borders because southern African countries were becoming independent in the mid-1970s. In response, the government embarked on a final push to realize its vision of regional development, which included independent African homelands. In this way, the extension of concerted governmental authority to the border region of Ellisras occurred in a messy procession, absent an underlying coherent design. A complex relationship of negotiation and compromise existed between Iscor and the various layers of government in the eventual enactment of forced removals.

3 ∾ Eskom and the Turning of the Tide

AS SOUTH Africa entered the 1980s, the National Party became doggedly determined to continue with the project of apartheid despite glaring red flags of its unviability. Verwoerd's vision of "grand apartheid" was burned alive as township residents engaged in continual protest action. Soweto had been in a state of perpetual unrest since the Soweto uprising of 1976, and in 1984 consistent rental increases and disagreement with their local councillors drove the normally quiescent residents of the Vaal Triangle to launch a campaign of strikes and stayaways.[1] The declaration of a state of emergency in 1985, which increased the power of the police to arrest and detain, only reinforced unrest. President P. W. Botha's declaration of a "total onslaught" in 1983 saw unbridled government spending on the military.[2] The end point was a crisis of government finances by the late 1980s—a fiscal crisis compounded by a stagnant economy, double-digit inflation, and the falling world price for gold. As fiscal precarity threatened the survival of the apartheid government, the state corporations persisted in expanding their operations, proceeding on the determinations of their demand forecasts.

Iscor had managed to see through the core components of its expansion plan in the 1970s, having built the Grootegeluk coal mine and its supportive infrastructure in the town of Ellisras. When Eskom entered the Waterberg in the 1980s to build the Matimba power station, it was in the midst of a program to expand its generating capacity, much like Iscor before it. This meant building six new power stations at an estimated cost of R175 million ($9.5 million) in total. Eskom issued its own bonds to raise money, though it struggled to attract foreign investors because of international hostility to apartheid.[3] President Richard Nixon's decision to leave the gold standard in 1971 had diminished South Africa's significance in the eyes of foreign investors and made them more likely to withhold investment dollars. This had occurred a couple of years before the oil crisis of 1973, demonstrating the general economic climate of straitened global financing.

Against this background, Eskom created a national electricity grid to attract foreign funding for its expansion plan, believing that the greater economies of scale of a centralized grid improved investor confidence.[4] To reduce its dependence on volatile inflows of foreign capital, Eskom harnessed domestic revenue sources by creating the Capital Development Fund (CDF), which functioned as a "sort of capital savings scheme"[5] into which it deposited 3 percent of its outstanding debt. Electricity tariffs increased for residential and business consumers in 1972, and in the first year of the CDF's existence, a 7.3 percent hike in electricity tariffs added R1 million ($54,000) to the fund. This was followed by a spectacular 20 percent increase in electricity tariffs each year for the next five years.[6] While the CDF was an important source of capital income for the expansion plan, the increased electricity tariffs elicited public consternation at Eskom's seemingly all-powerful position.[7]

Eskom named the project to develop the national grid the Central Generation Undertaking (CGU), and before its creation, the various regional authorities were responsible for distributing electricity to consumers within their region. The CGU created new geographies of electricity supply by linking all the power stations in the country together so that each power station contributed to a national electricity grid. Eskom built and planned a 400 kV national transmission grid, initially focused on connecting the Cape to the power stations in the Transvaal.[8] It allowed electricity to travel from the power station where it was generated to the regions of highest demand, even if the electricity was not consumed in the same region as the power station. While it is tempting to reduce the centralized electricity grid to a mechanism through which the government enabled the transmission of state power throughout the country, it is not clear that the government assumed the driving role in the project. But the government was on board with the initiative, and in Parliament, the minister of economic affairs defended

60 ～ *Apartheid's Leviathan*

Eskom's decision to centralize electricity provision by emphasizing its cost-saving virtues.[9]

Eskom had an installed capacity of nine thousand megawatts in 1971 and planned to add another ten thousand megawatts by 1980.[10] By 1973 Eskom had successfully centralized electricity provision on the national grid and adopted electronic management systems to improve its planning and coordination. It then embarked on the construction of its six new power stations, collectively dubbed the "six-pack," because they were completed in quick succession during the 1980s. The near-simultaneous construction allowed Eskom to reap the benefits of bulk buying from international suppliers, which meant that the power stations boasted equipment of a similar design. Most were built amid the coalfields of the Eastern Transvaal, known today as Mpumalanga. But Eskom could not situate one of its power stations in the Eastern Transvaal, due to the levels of air pollution in the region, and it was compelled to take up Iscor's offer of coal from its Grootegeluk coal mine. While unexpected challenges caused Eskom's engineers to stumble in the realization of their expansion plan, their tenacious efforts ultimately led them to join Iscor's operations in the Waterberg.

FORECASTING DEMAND

Eskom implemented the expansion plan on the basis of its estimations of the country's future demand for electricity. During the 1970s the country's electricity demand was the highest it had ever been, and Eskom thought its existing supply was precarious. Mozambican independence in 1975 had rendered South Africa's ability to access the electricity generated by the hydroelectric Cahora Bassa dam uncertain, though the dam ultimately proved a reliable source of electricity for South African consumers well into the postapartheid period.[11] South Africa had been assured access while Mozambique was under Portuguese colonial rule because Eskom had contributed to the construction and design of the dam, but it was unclear how receptive the Frelimo-led ruling party would be toward apartheid South Africa's continued use of its electricity. In addition, some of Eskom's own power stations, such as the newly built Koeberg nuclear power station, were experiencing technical and licensing difficulties, impinging on their ability to generate electricity.[12]

As the 1970s wore on, Eskom reassessed its electricity demand forecasts and the necessity of building all the new power stations set out in its expansion plan. In 1978, based on the input from regional administrations in different parts of the country of their future industrial and residential demand,

Eskom and the Turning of the Tide ⁓ 61

Eskom's management board revised its previous demand projections upward. In so doing, the board laid to rest debates over reducing the number of power stations in the pipeline, resolving instead to see the power stations to completion.[13] Eskom's demand forecasts also relied heavily on the fortunes of the gold mines, which were turbulent in the 1970s.[14] An expected increase in production at the gold mines loomed large in Eskom's decision to build six new power stations in the late 1970s. Following President Richard Nixon's decision to unpeg the US dollar from the price of gold in 1971, the world price for gold continued to rise steadily, and by 1980, the gold mines had not decreased their production.

In 1980 Eskom's forecasters again revised their demand forecast upward. Despite the uncertainty of the expected electricity demand of the Rand and the Orange Free State, both of which were centers of gold mining activity, they presumed that the gold mines would not reduce their demand for electricity in the near future.[15] This was because, even if gold mines reduced the amount of gold that was mined, miners would have to plumb greater depths and would have larger pumping loads. Electricity-driven pumps were required to pump groundwater out of the mines. Provided that the gold price remained sufficiently high, the forecasters expected that "production costs would also increase proportionally."[16] But the fortunes of gold fell sharply in 1979 when the US chairman of the Federal Reserve, Paul Volcker, raised interest rates in the United States to 20 percent, which lowered inflation and led to the decline in the price of gold. In the end, South African gold mines did not increase production as much as Eskom expected. In 1989 Mike Davis, Eskom's general manager for finance, told the *Financial Mail* that the sudden and unexpected collapse of commodity prices threw Eskom's load forecasts into disarray.[17] By the end of the 1980s, it was clear that Eskom had too much generating capacity. One of the newly constructed power stations, the Majuba power station, had to be mothballed when complete because Eskom had enough electricity generating capacity to satisfy the country's demand.

ESKOM'S RELUCTANT ENTRY INTO THE WATERBERG

In the early 1980s, while the going was still good, Eskom was proceeding full steam ahead with the construction of its six new power stations, which it planned to situate in the Eastern Transvaal. But the air pollution officer, whose office fell under the Department of Health, expressed his concern about the concentration of sulfur dioxide in the Eastern Transvaal. Sulfur dioxide is emitted during the coal combustion process and forms acid rain when mixed with moisture in the atmosphere. Acid rain then fell on the soil of White farmers in the region, and they complained of its effects on their agricultural produce. In 1977 the air pollution officer ruled definitively against the construction of one of the six new power stations, which Eskom had prematurely named Ilanga, in the Eastern Transvaal.[18]

Eskom first pleaded its case to continue with the power station's construction on the grounds that it would be difficult to find a replacement site, and it later tried to discount the air pollution officer's calculations by conducting investigations of its own into the sulfur content of the coal.[19] Its engineers also questioned the air pollution officer's methodology, leading him to admit that the calculations were not entirely conclusive, though his requirements were lenient when considered against the global norm. He remained convinced that the proposed new power station would raise the level of air pollution in the Eastern Transvaal to more than double the limit set by the World Health Organization.[20] The indeterminacy arose from the fact that it was only possible to measure the sulfur content of the coal, rather than the sulfur dioxide emitted during the actual coal combustion. In order to establish the latter, emissions would have to be measured under actual running conditions, which required the construction of forty testing stations at a prohibitive cost.[21]

At an Eskom board meeting held in January 1978, board members appreciated the necessity of securing coal supplies outside of the Eastern Transvaal but thought it best that Eskom refrain from mining the coal itself because the costs of coal mine exploitation would be heaped onto electricity tariffs.[22] In addition, Eskom would have to build connective infrastructure such as roads, rail, and telecommunications were it to exploit virgin coalfields. The Waterberg coalfields in particular were "unproven," which meant that the depth of the coal reserves and the ease of mining were uncertain, and board members thought that this mitigated long-term planning efforts. In addition, the Department of Defense had earmarked the Ellisras region as a security risk because of its proximity to the Botswana border, which meant that it was vulnerable to so-called terrorist activity from antiapartheid operatives housed across the border. The unsafe environment meant that Eskom would struggle to attract suitably skilled staff because they would have to live in a "confined environment in the power station township."[23] In this way, defense considerations made a power station in the Waterberg less attractive rather than more, and a station in this area would have conflicted with the government's imperative of preventing the depopulation of the border regions.

Eskom and the Turning of the Tide ⁓ 63

Despite these risks, Iscor's offer of coal from its Grootegeluk coal mine to supply a power station in the Waterberg did hold some appeal. This included signifying Eskom's solidarity with Iscor in serving the national interest and the financial advantage of sharing the costs of coal exploitation with Iscor. Then, in June 1978, Eskom was forced to concede to the air pollution officer after its own investigations concluded that the air pollution levels would indeed be intolerable if Ilanga were built in the Eastern Transvaal.[24] With no other option for the site of the power station, in August 1979 Eskom accepted Iscor's offer to supply it with coal from the Grootegluk coal mine and prepared to build its power station in the Waterberg. To coincide with the change of the site from Eastern Transvaal to the Waterberg, Eskom's engineers changed the name of the power station from Ilanga to Matimba so that the power station could better acculturate. As the Xitsonga word for "power" or "strength," they thought the name Matimba a more appropriate social fit for the languages spoken in the Waterberg.[25]

Environmental organizations were a source of some annoyance to systems planners in other parts of the world during this time as well. This is mentioned in a report from Eskom's chief engineer in systems planning, in which he detailed his visit to a meeting of the International Council on Large Electric Systems in Paris (CIGRE).[26] CIGRE was founded in 1921 as a meeting place for large-scale electricity producers from around the world.[27] The engineer's report on the conference proceedings details a distinct suspicion of environmental organizations (as well as trade unions) because they threatened the efficacy of long-term planning. He wrote: "It is clear from the experience in America and Australia that great care should be taken in South Africa to avoid the environmentalists becoming a major factor in planning. Possibly this can best be achieved by taking all reasonable steps to preserve the environment so that opposition was reduced to a minimum." Another delegate to the conference had reportedly said that "engineers must rebel against uneconomic and impractical standards." The Australian electricity utility represented at the meeting had been prevented from building a new power station where it had planned on the grounds that it would "wastefully use water." Trade unions prevented their construction of a natural gas power station, and public opposition to a nuclear power station meant that the utility had to rely chiefly on hydropowered electricity.[28] Thus, the systems planners, who relied on long-term planning stability, saw pressure from groups such as trade unions and environmental conservation organizations as best avoided rather than tolerated.

64 ~ *Apartheid's Leviathan*

THE MATIMBA POWER STATION

The Matimba power station was a high-risk proposition from the outset. The hot, arid, and water-scarce conditions of the Waterberg meant that Eskom's traditional wet-cooled power station design would not suffice; it would have to construct a water-saving, dry-cooled power station. But Eskom had only an experimental familiarity with dry-cooled systems, which meant that Matimba required an innovative design.[29] Nonetheless, Matimba benefited from a depressed global demand for power station equipment, and in May 1982 Eskom's board members reported that competition was high among engineering corporations for contracts to manufacture turbo generators and boilers.[30]

West German engineering corporations were particularly prominent in the South African market. For the Matimba power station in particular, the West German government had reportedly guaranteed R341 million ($18.6 million) worth of turbine and generator sales to Eskom. The West German government had decided to back export credit guarantees to South Africa in an effort to stimulate activity in the German engineering sector. This meant that the West German government carried the risk in case Eskom defaulted on its payments for the equipment.[31] A depressed domestic engineering sector drove West Germany's amenability to South Africa, despite the growing international campaign for disinvestment. Its engineering corporations had been hard hit by the indefinite postponement of the construction of nuclear power stations as the antinuclear party, the Green Party, gained political ground in the country. The West German government had also secretly collaborated with South Africa to develop its uranium enrichment facilities and build nuclear weapons, intelligence that the African National Congress seized upon in 1975 as part of the international antiapartheid campaign. As Gabrielle Hecht and Paul Edwards write, antiapartheid activists and scholars subsequently argued that the transnational collaboration sustained apartheid technology and expertise and, ultimately, apartheid itself.[32]

Eskom's ambition for Matimba was unmatched in its power station fleet. Its engineers designed the power station with the region's environmental conditions and security considerations in mind, and the result was a power station with novel technological features that they had little experience handling. A special edition of the magazine *Engineering Week,* which appeared in November 1987, celebrated Matimba's engineering feats.[33] The special edition featured interviews with the leading engineers responsible for the construction of the power station. While the articles in its pages

Eskom and the Turning of the Tide ⟿ 65

lacked any critical engagement with the activities of these engineers, they provide valuable insight into the design and construction of the power station. *Engineering Week* celebrated Matimba's world-class technological features, which made the power station "yet another 'world first' for South Africa."[34] Alec Ham, Eskom's general manager of engineering who oversaw the completion of Matimba, praised the engineering team at Matimba for having "converted an untried concept into what has become a world leader in dry cooling for power generation."[35] Matimba's site manager, John Begg, said that on completion the power station would have a four-thousand-megawatt generation capacity, making it the largest power station in the southern hemisphere.[36]

Cooling systems are crucial features of coal-fired power stations, designed to cool the large quantities of steam generated during the coal combustion process. Coal combustion generates heat, which evaporates the water in the boilers; the resultant steam is forced through the boiler pipes to drive the turbines, which in turn generate electrical energy. Thereafter, the cooling system condenses the steam, and the water formed through condensation is recycled and returned to the boilers. Ham said that in the absence of a cooling mechanism, the heat emitted is comparable in quantity to that of a massive forest fire. While Eskom's typical wet-cooled condensers were cheaper than dry-cooled systems, they only made sense when water was abundant. A dry-cooled system avoided a reliance on water by using enormous fans powered by a portion of the electricity generated by the power station, which came with its own disadvantage of electricity waste.[37]

Eskom had little to no experience constructing a dry-cooled power station that was Matimba's envisioned size. In fact, nowhere in the world was one in existence. The closest comparable power station—with a generating capacity not nearly as large as Matimba's projected capacity—was a 375-megawatt direct dry-cooled unit that formed a part of the Wyodak power station in Wyoming in the United States. Other examples of dry-cooled power stations were the van Eck power station in Namibia (then called South West Africa) and the Utrillas power station in Spain, both constructed by the German-based engineering company GEA. Eskom had conducted its own experiments into the feasibility of dry-cooling technology, though this was at power stations with a limited generating capacity. One of these, the Grootvlei power station, situated near the small town of Balfour in the Eastern Transvaal, boasted two dry-cooled units, each with a capacity of two hundred megawatts. These units used the natural or "indirect" method of dry cooling, where steam is cooled by a natural draft of

66 ～ *Apartheid's Leviathan*

air that forms within a large cooling tower. These cooling towers were necessarily tall to effectively cool the steam, relying on the principle that the warm, moist air would rise and would cool as it rose, so creating a constant draft. This method of cooling differs from the "direct" dry-cooled system where large motor-driven fans generate a draft large enough to cool the hot steam that passed through the turbines. The successful operation of the dry-cooled units at Grootvlei inspired Eskom's engineers to expand the generating capacity of its dry-cooled power stations.[38] In April 1980 they detailed the findings of new experiments undertaken by overseas turbine manufacturers and stated their conviction that it was possible to build a six-hundred-megawatt dry-cooled unit, though it would cost 7 percent more than a wet-cooled unit.[39]

In 1982 Eskom decided that the Matimba power station would have a direct dry-cooled system, which required less capital but had higher operating costs. Since the cost of coal made up a large proportion of the operating cost estimates, Eskom's decision to opt for a direct dry-cooled system meant that it favored higher operating costs over capital costs. This was likely made feasible by the low cost of Waterberg coal, although this rationale is not explicitly stated in the records. Defense considerations and the vulnerable position of the power station, near as it was to the Botswana border, ultimately proved decisive in Eskom's choice of a direct dry-cooling system as opposed to an indirect one. The tall cooling towers required for indirect dry cooling would be easy targets for terrorist activity in the stark bushveld, and the reconstruction of a damaged tower would mean cost escalation and delays in completion. Smaller cooling towers were preferable, but these were more expensive to build.[40] A direct dry-cooled system, however, did not require the construction of tall cooling towers and so reduced Eskom's presumed vulnerability to sabotage.

Matimba was built during a boom in the South African construction industry. In the 1980s the long-standing threat of sanctions against South Africa had become a reality, and domestic manufacturers reaped the profits of the country's global isolation. In addition, the depreciation of the rand since 1985 made imports expensive, and South African manufacturers stepped up to the plate to meet the demands of large industrial projects across the country. One engineer who worked at the Matimba power station construction site during the 1980s described the decade as the "golden age of power generation."[41] His colleague echoed this sentiment: "I think if you go back to the time when Matimba was being built, the boiler industry in South Africa was really at its peak. We had four boiler

Eskom and the Turning of the Tide ⁓ 67

companies that had their own facilities where things were getting done. The industry was really at its height."[42]

President P. W. Botha, consumed with his "total strategy," encouraged large infrastructural developments to ensure the country's self-sufficiency. But another consequence of this government-directed spending was the rapid transfer of capital from the public to the private sector. The government compelled Eskom to support local manufacturers by awarding them equipment supply contracts, and global engineering companies who had incorporated domestic manufacturers into their operations were looked upon favorably. As a result of the continual power station construction projects, there existed a tight circle of cooperating corporations, made up of few global specialized engineering corporations and South African manufacturers.

During deliberations among Eskom's engineers about the preferred supplier of the turbines for Matimba, it emerged that the turbine posed the greatest design challenge for the dry-cooled power station. The turbine contained rows of blades that rotated as the fast-moving steam passed over it, and the blades in the back row of the turbine needed to withstand a higher back pressure than in a wet-cooled system. Since wet-cooled power stations were more commonly used globally, few engineering companies had a proven track record for supplying turbines for dry-cooled power stations. Eskom's engineers visited the plants of the different turbine manufacturers in Europe and Japan, including Maschinenfabrik Augsburg-Nürnberg (MAN) in Germany, Brown Boveri in Switzerland, General Electric Company in the UK, and Toshiba, Hitachi, and Mitsubishi in Japan. They reported that of these, MAN was the only manufacturer who had tested its turbine design under actual running conditions and could proclaim its confidence in the turbine's ability to withstand the higher pressures.[43] MAN won the tender for supplying the turbine along with the French-based Alsthom.[44] MAN was also attractive because it had prior experience with Eskom, having supplied two turbogenerators to the Grootvlei power station. To enable coordinated planning between the different corporations involved in the construction of the power station, MAN had initiated talks with the boiler contractors with the view to designing an all-encompassing control and instrumentation system that integrated the boiler and turbine generator operations. Communication between the two companies was helped by the fact that MAN owned a part of EVT, one of the corporations that held the contract to build the boilers.[45]

The boilers were another crucial and expensive component of the power station. Local subsidiaries of the global companies responsible for

68 ⁓ *Apartheid's Leviathan*

supplying the power station's specialized engineering equipment were represented by a company called Industrial Machinery Supplies (IMS). IMS had a long-standing relationship with Eskom, having first supplied the steam turbines for the Komati power station in 1961. The boiler contracts were eventually awarded to a consortium, made up of the German engineering firms Steinmüller and EVT, that called itself Sieva. IMS held a 6 percent stake in its operations, and the fact that the consortium included a South African company made it more attractive to Eskom because the inclusion of local manufacturers was an explicit component of Eskom's tender specifications.[46] In 1989 the South African weekly *Financial Mail* detailed IMS's spectacular rise in fortunes. During the preceding five years, its annual turnover had increased from R30 million ($1.6 million) to R150 million ($8.1 million), highlighting the role of capital transfer in the South African economy through Eskom's power station construction.[47]

Sieva assumed overall responsibility for the boiler installation and coordinated the subcontractors responsible for the mechanical and civil works. While Sieva was a consortium of largely foreign firms, it gradually adopted a local staff complexion by incorporating South African engineers into its ranks. During the boiler design phase, the consortium appointed a project supervisor to coordinate the different companies; he was based in France until 1986, when Sieva's personnel assumed the role in South Africa. The consortium boasted that one of its main contributions to local manufacturing was the creation of a company called the Boiler Component Manufacturers of South Africa (BCM) in 1982 to manufacture boiler pressure parts. The remainder of the construction work was subcontracted to various specialist companies, including LTA for the civil works, Genrec for steelwork, and Steinmüller for the erection of boiler pressure parts and pipework.[48] This was in large part to spread the financial benefits related to the construction of a large infrastructure project such as Matimba.

MATIMBA AND ELLISRAS

Having committed to the Waterberg coalfields, Eskom had to engage with the imperatives of socioeconomic development in the town of Ellisras, working alongside intransigent local government authorities who preferred that Iscor and Eskom pay for new infrastructure themselves. This was the case with the construction of a hospital that could treat both African and White patients. Eskom was not legally permitted to operate the hospital because it was not a core part of its operations, but it agreed to share the estimated cost of R1.7 million ($92,000) with Iscor on the condition that the

Eskom and the Turning of the Tide ⟶ 69

provincial government eventually assume responsibility for the operation of the hospital, preferably on a lease agreement so that Eskom could gradually recover its capital expenditure.[49]

In the later part of the 1980s, Eskom was instrumental in the formation of the African township of Marapong, located close to Matimba and the Grootegeluk coal mine. This echoed the gradual loosening of the strictly enforced mandates of the Group Areas Board, which was taking place all across the country at the time. In 1979 a commission of inquiry into influx control, called the Riekert Commission of Inquiry, had encouraged the development of permanent African residential neighborhoods in urban areas and the development of an African middle class. In some ways, this marked the resurgence of the debate over stabilization policies in African cities that had occurred during the period leading up to the Second World War. British colonial officials had adopted the term "stabilization" as a euphemism for recognizing the permanency of an African urban population in response to waves of worker unrest that swept their colonies during the 1930s. Fred Cooper describes stabilization as a system of "separation"[50] that distinguished between the rural poor and the urban African working class. Similarly, critics of the Riekert Commission's recommendations argued that it created a fenced-off minority of prosperous Africans in nominally White urban areas that would function as a "buffer" against legitimate protest from the impoverished residents of the homelands.[51]

Eskom adapted to the imperatives of this new era, within which the older order of racial segregation was being dismantled. In 1980 a memorandum written by its personnel manager concluded that there was an urgent need to provide housing for married African workers.[52] While this would be expensive for Eskom, the personnel manager thought that the costs could be offset by a "stable and content" workforce. He also emphasized the importance of Eskom's decision to review its housing policy in light of the recommendations of the Riekert Commission of Inquiry of 1979.[53] Eskom's engineers considered apartheid's segregationist dictates to be inimical to the development of a suitable African workforce. During the construction of the Matimba power station, Eskom began to consider the possibility of encouraging mixed-race neighborhoods in the vicinity of its power stations at a national level.

Writing in 1986, the Matimba power station manager estimated that the power station required eleven hundred permanent African workers once it was operational and five thousand temporary workers during the construction period.[54] Matimba's White workforce amounted to six

70 ～ *Apartheid's Leviathan*

hundred families, who were to be housed in Onverwacht alongside Iscor's White labor contingent. But African workers were subject to a different housing regime and were barred from Eskom-owned family quarters in line with government regulation. The prevailing migrant labor system dictated that only men would be accommodated in the Ellisras district, while their families remained in the homeland to which they were relocated. During the construction of the Matimba power station, only 3 percent of married Africans could legally be housed in the region of the power station. The chief regional African township, Mokerong, was situated thirty miles (fifty kilometers) away from Ellisras, within the borders of the Lebowa homeland, which made daily transport to and from Matimba expensive. Highlighting the difficulty of transport, Eskom argued for the need to establish an African township closer to Ellisras because it expected to employ twelve hundred African workers by the year 1991.[55]

Iscor was at first reluctant to support Eskom's bid to establish an African township. Iscor's general manager of open-cast mines, J. P. Deetlefs, objected on the grounds that the construction of an African township was not in the immediate interests of the coal mine and that Iscor was unwilling to break government legislation on housing segregation.[56] In addition, Iscor would have to carry some of the capital costs for family housing when, Deetlefs argued, it already owned enough houses, property, and facilities to service its own labor force. Importantly, the development of an African township defied the sentiment of the townsfolk and risked alienating influential White members of the Ellisras town. In September 1987, community newspaper *Die Kwevoel* detailed the conflict between the various constituencies of Ellisras.[57] The main body representing famers in the district, the District Agricultural Union (DLU), began a petition signed by its own members and various town residents to protest the development of the township. The DLU was concerned that there would not be enough water to go around for the township residents, and it argued that the addition of land to the homelands prevented the need for an African township in Ellisras.[58] Half of the town's registered voters had reportedly signed the petition that was to be handed to the minister of agriculture. While the political parties, the Herstigte Nasionale Party and the Conservative Party, had already voiced their protest, the DLU was particularly proud that their petition represented the voices of residents across political party lines. Andries Treurnicht, the founding member of the Conservative Party and the Waterberg representative in Parliament, also set out his clear opposition to the project in a memorandum to government.[59] Finally, the *Mogol Kommando,* a decades-old

Eskom and the Turning of the Tide ⁓ 71

quasi-military body, argued that migrant labor was a common labor practice across the world and that a permanent African population in White urban areas would only worsen the labor shortages in rural districts.[60] Rather than creating a township, they argued, the government should focus on improving the bus service between the homelands and Ellisras.

But the opposition proved futile and the state corporations had their way in the end. At the time the DLU submitted the petition, Iscor, Eskom, the town council, and the two chambers of commerce in Ellisras had already agreed in principle to proceed with the development of the township. Eskom applied for special permission to President F. W. de Klerk in 1987. De Klerk then wrote to Chris Heunis, the minister of constitutional development and planning, and after consultation with members of Parliament, the creation of the township was approved. Having gained special permission from the president, Eskom and Iscor proceeded with its development in direct defiance of the wishes of many of the White residents of the town.[61] They thus initiated the development of family housing for African workers in the Marapong township.

The construction of the Matimba power station represents the last hurrah of Eskom's engineers as systems planners; they had proceeded tenaciously in the realization of their generation-capacity expansion plan. Eskom built its six new power stations in the 1980s amid a state of general unrest in the country, combined with the growing incoherence of the apartheid project. Despite the scarcity of funding and the unexpected opposition it encountered from the air pollution officer, Eskom dug in its heels and saw its expansion plan to completion. The centralized, national grid as well as Matimba's unprecedented design that allowed it to survive the new, arid conditions of the Waterberg arose as solutions to these problems. Coincidentally, the amenability of West German electrical engineering firms to supply equipment to South Africa was a product of opposition to the development of nuclear power plants from German environmental conservation groups. At the end of the 1980s, Eskom oversaw the development of an African township close to the Matimba power station. This occurred soon before the final expiration of the precepts of the Group Areas Act, and in the face of substantial opposition from residents of Ellisras, Eskom had to acquire special permission from newly installed President De Klerk. This set the scene for Eskom's subsequent efforts to adapt to the imperatives of the new era of democracy and encourage racial integration at Matimba.

72 ～ *Apartheid's Leviathan*

4 ~ Contested Neoliberalism

AS THE 1980s drew to a close, the crisis of state finances became apparent to the apartheid government. The fall of the gold price in the 1980s, combined with unfettered military spending, created a heavily indebted government.[1] For international creditors, the continued unrest in the townships denoted South Africa's interminable political instability, and in 1985 certain US-based banks placed a moratorium on further loans to the South African government. This marked the onset of what can broadly be termed neoliberalism in South Africa nearly a decade after it took hold in European countries. While inchoate in many respects, the term denotes a point in time in which countries around the world began relying less on government spending (and the government paternalism this entailed) to embrace the neoliberal economic orthodoxy. This orthodoxy is generally associated with a group of thinkers who belonged to the Mont Pelerin Society, named after the Swiss mountain resort where they first met in 1947. The ascendency of Keynesianism after the Second World War had relegated these neoliberal thinkers to the margins of economic orthodoxy until the 1970s, when European economies confronted a crisis of stagflation. With slow economic growth, high rates of unemployment, and high rates of inflation, stagflation prompted a radical revision of economic policy.

In South Africa, as all around the world, fiscal crisis prompted reform, and reformers targeted the reduction of what they viewed as excessive levels of government spending. Across the African continent in the 1980s, the World Bank and the International Monetary Fund introduced structural adjustment policies (SAPs), the presumed capsule of neoliberalism in Africa. Scholars such as Fred Cooper have suggested that SAPs marked the end of the postcolonial developmental era by diminishing the government's capacity to deliver public services. In return for life-giving loans, the SAPs prescribed the sale of state assets, the diminution of the civil service, and the removal of barriers to entry for imported goods. These measures caused little to no economic growth, however, with only Ghana and Uganda achieving a modicum of long-term economic success.[2] But Nicolas van de Walle questions the extent to which SAPs were actually adopted, suggesting that they were hardly obeyed to the letter. In many countries, the civil service remained bloated,[3] and in general, sketchy data availability makes it difficult to draw a convincing image of the actual effect of SAPs.[4] In this vein, rather than viewing neoliberalism as a monolithic, all-encompassing entity, this chapter describes its weakened and compromised adaptation. In the case of Eskom, despite the enthusiasm of reformers who advocated views in accordance with neoliberal orthodoxy—the importance of individual economic freedom and of the value of competition[5]—its continued developmental role and relationship with the government remained paramount.

Neoliberal reform in South Africa in the 1980s was not merely a foreign imposition but fitted instead onto preexisting domestic configurations of power. The composition of Afrikaner nationalism had irreversibly altered over the preceding decades, resulting in a factional split between the *verkramptes* and the *verligtes*.[6] The verligte camp grew in strength after the Soweto uprising of 1976, setting the stage for the eventual relaxation of racial segregation and the inability of the apartheid government to penetrate as deeply into the lives of its citizens as it once could.[7] Antina von Schnitzler argues that, following the Soweto uprising, government-linked economists assimilated the writings of neoliberal thinkers into their own policies in a bid to ensure the viability of White minority rule. They were particularly attracted to the neoliberal conception of the free market as something that required government intervention to bring into being.[8] Von Schnitzler describes the work of an influential South African economist, Jan Lombard, who sought the reform, rather than the dissolution, of apartheid in line with the verligte ethos, and who served as the deputy governor of the South African Reserve Bank for a time. In Lombard's two books on the subject, published in 1978 and 1979, he recommended that Africans should be subject to a regime of "benevolent paternalism"

74 ⁓ *Apartheid's Leviathan*

to ensure that individual behavior could conform to economic freedom within an eventual liberal order.[9]

Economic freedom necessitated cultivating the ethos of individual capital accumulation. Although the government had previously removed capital from the hands of Africans, during the 1980s government officials worked to create a group of African entrepreneurs in the townships and in the homelands. While this was successful to a degree, it was not by any measure a far-reaching initiative and it benefited only a tiny minority. During this period of apartheid reform, neoliberal precepts were selectively appropriated in a top-down manner to make apartheid more palatable to the general populace. Africans could remain politically "rightless" for an indefinite period while economic differentiation slowly superseded racial segregation. By the time of the financial crisis of the late 1980s, the National Party (NP) had moved decisively toward abandoning the poor Whites it had previously endeavored to assist.[10] But by then, arguably, the process of transferring capital from the government to uplift the economic condition of the White citizenry had already run its course.

SOUTH AFRICA'S DEBT CRISIS OF 1985

In 1985, a debt crisis made reform especially urgent. Since the oil crisis of 1973, international creditors had curtailed their loans. But a boom in the gold price between 1978 and 1980 brought in sufficient foreign exchange earnings to tide the country over and allow the government to continue purchasing oil and arms. Then, in 1981, both the public and private sector went on what Vishnu Padayachee has termed "an orgy of borrowing from private international banks directly and from the international capital market via bond issues."[11] While South Africa's credit rating had improved in the aftermath of the 1976 Soweto uprising, banks in the US watched developments in the country nervously as mass unrest in the townships gathered force in 1984.

On 20 July 1985, President P. W. Botha declared a state of emergency, and a few days later, on 1 August, the New York–based Chase Manhattan Bank placed a moratorium on all further loans to South Africa, precipitating the cessation of loans from other US banks. In response, the South African government announced that it would be unable to make repayments on its outstanding debts until March 1986. South Africa had also relied heavily on the loans of a cluster of West German and Swiss banks, and by 1985 even these once reliable funders had cut off South Africa.

Eskom was not immune to the debt crisis, despite the fact that it could issue its own bonds independent of the government. During 1984 Eskom,

Contested Neoliberalism ⁓ 75

in the midst of its power station building spree, had become the country's largest borrower of medium-term funds, having secured a R50 million ($2.7 million) loan without the guarantee of the South African government, a rare achievement for its time.[12] But in July 1985 it confronted a crushing lack of foreign investor enthusiasm for a bond issue, in stark contrast to the unruffled investor confidence it had enjoyed only a year earlier. In 1986, having failed to borrow the R1,400 million ($76 million) it had hoped, it was forced to cut R1,000 million ($54 million) in operational expenditure and R1,000 million in capital expenditure from the amount it expected to spend between 1986 and 1989.[13]

While the financial crisis provided the incentive for the privatization of state corporations, privatization was not an inevitable outcome of austerity.[14] The drying up of foreign loans created a context of crisis that was most conducive to fundamental reform. In many other African counties, for instance, austerity measures were part of the loan conditionality contained within the controversial SAPs. There is little to suggest that these international monetary institutions played a similar role in South Africa, however.[15] It is also not clear that the government was compelled to adopt privatization measures to appease foreign creditors. Most accounts of Western leaders' interventions in South Africa during this period describe their abhorrence for apartheid and the pressure they put on the NP to institute democratic reforms. While Margaret Thatcher promoted privatization in the United Kingdom, she did not seem to impose the idea and process on the South African government. She had come to be a reliable, though often chiding, international ally for the NP and held off imposing sanctions against South Africa for as long as possible.[16] If Thatcher encouraged greater private sector involvement in the economy, it is clear that she left the details of this process up to the South African government.

In 1987 the government released a white paper on privatization and deregulation[17] that criticized the high levels of public spending in South Africa in the preceding decades. Public spending had increased substantially since 1975 and now constituted the bulk of national spending.[18] Much of this went toward the government's project of encouraging acquiescence through increasing services to Black townships and implementing industrial decentralization initiatives to provide employment to homeland residents. In May 1990, the minister of public enterprises announced in Parliament that the government did not see the privatization of state-owned corporations as a necessary end point, and that market efficiency could flow from "commercialization" rather than a simple transfer of ownership.[19] The South African government was thus chiefly concerned with resolving its debt crisis, and it was prepared to do so without the sale of state assets.

WIM DE VILLIERS AND PRIVATIZATION

To evaluate the feasibility and practicalities of privatization, the government created a ministerial committee for privatization and deregulation, headed by a longtime critic of the state corporation model, Wim de Villiers. Having long proclaimed his disdain for the low-wage paternalist system that dominated the South African economy of the twentieth century, de Villiers was particularly critical of the peculiar structural foundation of South African state corporations. "A strange sort of organization has emerged," he said in an interview with the *Business Day*. "It is a functional organization that does not divide into business units, demands no profits and delegates no powers. Its blind concentration on function has led its management to spend freely on the most technically advanced equipment, irrespective of the capital cost or the resultant output."[20] De Villiers also expressed views in line with neoliberal orthodoxy, such as the need to reduce the size of the civil service.[21]

An engineer by training, de Villiers had managed a Zambian copper mine owned by the South African mining corporation Anglo-American in the 1950s. He returned to South Africa in 1961, where he eventually became the chief executive officer of General Mining Union Corporation (Gencor), which he transformed into a centerpiece of Afrikaner commercial efficiency.[22] He subsequently served as the director of the defense equipment manufacturing corporation Atlas, where he was reputedly successful in "reinforcing realistic manufacturing capabilities."[23] By the time of his appointment to the Committee for Privatisation and Deregulation, de Villiers had earned the reputation of "making capital work"[24] and introducing the commercial imperative to sluggish, Afrikaner-linked corporations.

De Villiers's report on the privatization of Eskom in particular was published in 1984.[25] His recommendations were implemented without any apparent opposition, although they were not dramatic enough to arouse any indignation. He recommended not the privatization of Eskom, but the creation of a two-tiered management board to oversee Eskom's commercial restructuring. He was reportedly inspired by his experience at the Anglo-American Rokahana copper mine in Zambia, where he concluded that Africa's economic growth was constrained by a shortage of management skills and capacity. On further investigation, he came across an executive

Contested Neoliberalism ⁓ 77

management model then operative in certain European companies (particularly West Germany) that adopted a two-tier management structure. When put into practice, one of the tiers contained a nonexecutive board made up of "consumer-interest directors" who oversaw the operations of the management board. The management board, in turn, oversaw the day-to-day operations at Eskom.[26] Eskom retained its status as a statutory body, a status that was legally revoked only in 2001 when it was transformed into a public company. With the additional layer of management oversight, the government could point to the reining in of Eskom's managers and engineers and their greater accountability to the public.

In a report released on 25 May 1988, de Villiers detailed his views on the privatization of Eskom more clearly, along with the question of the privatization of the South African transport services and the post and telecommunication services.[27] Above all, he was concerned with preventing the formation of "natural monopolies," a term used to describe a sector or corporation that is especially reliant on technology. For these generally networked infrastructures, the rationale goes, the capital investment is so high as to bar entry to new arrivals. De Villiers warned that a privatized state corporation could continue to dominate the market and thus constitute a private monopoly. Because of the state corporations' decades-long dominance of their respective sectors and their accumulation of capital and expertise over the years, new competitors had little chance of survival. This view was not without its critics. Frank Verhies, a lecturer in business economics at the University of the Witwatersrand, opined in the *Financial Times* in November 1990 that "Wim de Villiers is a big fan of 'natural monopolies.' He used this hoary old myth last year to recommend against allowing competition for the telephone network. So we're stuck with bad service, scratchy lines and outrageously high overseas charges."[28]

De Villiers drew widely and exclusively on the British experience with privatization as a basis for comparison. He argued that the South African energy market differed fundamentally from its British counterpart because there were no preexisting electricity providers that could offer consumers an alternative to the national electricity utility. When Britain passed its Electricity Act of 1989, which allowed for the privatization of components of the national electricity grid, de Villiers told the *Business Day:* "In Britain we are looking at industries that used to compete but were subsequently nationalised. In this country we are looking at monopolies."[29] He thought that Eskom would likely become a natural monopoly if it was privatized, thereby creating a noncompetitive market.

78 ⁓ *Apartheid's Leviathan*

In May 1990 the minister of public enterprises told Parliament that while the privatization of Eskom was still technically under review, the government was not convinced it was the right decision. He praised the transformation wrought by the two-tier board because it ensured the presence of a "strong policy forming and supervisory council" staffed with businessmen experienced in management.[30] Importantly, the minister acknowledged Eskom's diplomatic role in southern Africa since it had extended its electricity grid into neighboring southern African countries in the hopes of creating a network "which [would] stretch from the equator right down to the Cape."[31] While the minister supposed that Eskom had brought together the warring parties of Renamo and Frelimo in neighboring Mozambique, in reality, as Allen and Barbara Isaacman have shown, the apartheid regime used Mozambique's electricity supply to exert pressure on the Frelimo-led government. The apartheid government had nurtured Renamo from its inception and encouraged its sabotage of the electricity pylons in Mozambique, so harming President Samora Machel's hopes of economic transformation after colonial rule. The Nkomati Accord of 1984, which concluded negotiations between Renamo and Frelimo, stipulated that the Frelimo-led government would close down African National Congress (ANC) bases in Mozambique in return for Renamo's pledge to halt the destruction of electricity infrastructure.[32] In the end, the South African government valued Eskom's diplomatic power in southern Africa too deeply to sanction its dissociation.

The impending political transition of the early 1990s influenced the outcome of the privatization efforts. When President Botha left office in 1989, his successor, F. W. de Klerk, began negotiations with the exiled ANC, and this took the question of privatizing Eskom off the table.[33] Both the ANC and the Congress of South African Trade Unions (COSATU) called for Eskom to remain in public hands, and its privatization would have been viewed as an act of bad faith during the negotiations.[34]

The idea of Eskom remaining a state-owned entity was made more palatable to the public by its improved financial fortunes, mainly because it had reduced its spending in the late 1980s. Furthermore, the expected inflation rate for 1992 was 14 percent, but Eskom announced a lower increase in electricity prices—only 9 percent, which assisted an embattled government in its efforts to reduce inflation rates.[35] But lower electricity tariffs, argues Anton Eberhard, were more a product of capital investment than of greater operational efficiency. Eskom's lack of capital investment continued into the ensuing decade, and between 1992 and 1997, its revenue exceeded its capital requirements.[36]

Contested Neoliberalism ∼ 79

Iscor, however, was privatized in 1989 and its shares were made public on the Johannesburg Stock Exchange. In 1988 de Villiers told a news publication that Iscor was suitable for privatization because it had proven that it was able to operate at a profit, having "been subject to market discipline for a long time."[37] This market discipline arose in part from its experience with raising money on the international market. In 2001, Iscor was unbundled, which meant that its steel manufacturing plants were separated from its mining components. It remained a steel manufacturer under the name Iscor and its mines were sold to a company called Kumba Resources. In 2004 Iscor was bought by Mittal Steel, owned by Indian steel magnate Lakshmi Mittal, who in 2006 formed Arcelor Mittal, the largest steel manufacturing company in the world.

THE PRIVATIZATION DEBATE OF THE TWENTY-FIRST CENTURY

The question of privatizing Eskom fell under the spotlight again during the presidency of Thabo Mbeki, who had served as deputy president since 1994 and became president in 1999, when President Nelson Mandela left office. As deputy president, Mbeki oversaw the ANC's abrupt turn from its Reconstruction and Development Programme as long-term policy for economic development and its shift instead to the Growth, Employment, and Redistribution plan in 1996, heralding the ruling party's neoliberal embrace.[38] In contrast to the concerns about inefficiency and wastage that drove the privatization process of the late 1980s, this resurgent attention emanated from a small group of technocrats within the cabinet. These advocates of privatization, as Anton Eberhard writes, were "observing international trends in power sector reform, and were beginning to be concerned with the potential problems of monopoly power."[39]

In 1998 the government released its white paper on the supply of electricity in the country, which advocated a departure from the supply-heavy preoccupation of past energy policy.[40] The institutional memory of the erroneous demand forecasting of the 1980s was still fresh. In line with neoliberal belief in the impossibility of a monopoly provider accurately predicting demand, the white paper emphasized the importance of demand-side determinants of energy provision.[41] The white paper was also concerned with reviewing the country's energy policy to better fit the dictates of governance in a democratic society, which meant gearing electricity supply toward better service for the majority of households in the country, especially poor households, and away from its traditional focus on the mines and heavy industry. Amid the renewed uncertainty over the privatization of Eskom,

80 ⁓ *Apartheid's Leviathan*

the government discouraged spending any more funds on new electricity generation capacity. This decision led directly to Eskom's supply crisis a decade later.[42] In 2007, President Mbeki apologized for load shedding at an ANC fundraising dinner in Bloemfontein, declaring that "Eskom was right and government was wrong."[43] The government had been preoccupied with diluting Eskom's monopoly control of the energy market.

Following the government's return to the question of privatization, certain senior managers at Eskom were tasked with investigating the viability of an entirely privatized Eskom.[44] Clive Le Roux, an engineer at Eskom who served as the power station manager at Matimba in the early 1990s, was part of a team that studied the European Union's then current rules for competitive industries. Le Roux had visited Germany in the immediate aftermath of the fall of the Berlin Wall and visited both East and West Berlin. He described the experience of passing through the Brandenburg gate as akin to "walking over a line on the earth where there were forests and greenery and birds singing and suddenly there was death and broken tree stumps and grey—no color."[45] This experience highlighted for him the ravages of the state-planned economy.

Le Roux's team found that for Eskom to be privatized but avoid remaining a monopoly, it would have to conform to European Union regulations on market competition. A market was deemed competitive on the basis of the number of corporate entities within it. The more entities in existence, the lower the market share of each, and the more competitive the market in its entirety. The team discovered that for a privatized Eskom to conform to these requirements, it would have to split into seven different business entities on the basis of the different functions that Eskom performs (such as generation, transmission, distribution, etc.), each of which would constitute a discrete corporation. This effort to create a privatized electricity sector was ultimately unsuccessful—no investor was willing to buy the utility, or at least portions of it, and then produce electricity at the low tariffs required to hasten mass electrification of households in the country. According to Le Roux: "The price was too low; nobody would buy them and the whole thing was scrapped."[46]

In August 2001, the government passed the Eskom Conversion Bill, which ensured that Eskom fell under the provisions of the Companies Act of 1973.[47] This transformed Eskom into a public company with shares owned by the state. COSATU had voiced its objection to the Conversion Act before its passage, arguing that it paved the way for the privatization of Eskom. In his submissions to the Parliamentary Monitoring Group detailing its objection,

Contested Neoliberalism ⌢ 81

COSATU's assistant general secretary argued that privatization would increase the costs of electricity and that "it was Cosatu's great concern that the question of electrification should remain in the hands of the electorate and be governed by it."[48] Eskom was not taken further down the road to privatization thereafter. Eberhard has suggested that this was in part due to the absence of strong political will to unbundle Eskom and proceed with actual privatization.[49] In addition, the strong resistance from COSATU, which was a member of the governing tripartite alliance, would also have contributed to rendering privatization unlikely.

MATIMBA'S NEW REGIME

While the government toyed with the idea of privatization at the national level in the 1990s, Eskom, too, was allured by marketization and the promise of commercial efficiency in its inner workings. In 1985, John Maree had replaced Jan Smith as the chairman of Eskom. In a break with the traditional qualifications of Eskom's chairmen, Maree was not an engineer by training but held a degree in commerce from the University of the Witwatersrand. As Le Roux argues, when Maree took over, Eskom changed from an "engineering organization" to a "business."[50] Le Roux arrived in Ellisras in 1992 with a mandate from Eskom's head office at Megawatt Park to reform management practices at the Matimba power station. In line with its commitment to commercial reform, Eskom had introduced a decentralized cost-revenue accounting method across its power stations. In this model, power stations were treated as individual business units with their own annual financial statements detailing their costs and revenue. The annual revenue of a power station was calculated on the basis of the volume of electricity sold multiplied by the price of transfer. Le Roux considered the model an attractive method of controlling costs because it allowed the power station manager to zero in on the sites of wastage.[51] Part of Le Roux's mission was to correct mechanical faults that were impeding Matimba's productivity.

While Matimba was then one of the newest power stations in Eskom's fleet, and much like the Medupi power station, utilized cutting-edge technology for its time, its actual performance was below par. Together with the head of engineering at Matimba, Le Roux oversaw the replacement of various failing components of the power station. This involved halting the power station's operations and taking the plant off-line so that the necessary repairs could be made. In cases where serious and costly repairs were required, the contractors (in this case the consortium known as Sieva)

82 ⌒ *Apartheid's Leviathan*

refused to accept responsibility. But in the new climate of cost saving, Le Roux could not accept this without a fight.

However, it remained difficult to determine the root causes of technical failure and to ascertain which party had to bear the blame. In the absence of an irrefutable diagnosis of the technical fault, engineers from the contracting company and from Eskom alike disagreed on the reasons for failure. Despite investigations into the causes of technical breakdown, there remained a substantial degree of indeterminacy that had to be negotiated by the interested parties. Following years of litigation, Le Roux concluded that conflict with contractors almost always ended with both parties sharing the blame and carrying half of the cost of repair or replacement. This was because the court cases rested on competing expert analyses that lay beyond the courts' ability to resolve on a strictly legal basis. In this way, the imperative to save costs opened the door to a seemingly irresolvable technical uncertainty.

One of the most litigious conflicts involved a crack that surreptitiously appeared in the boiler exhaust ducts. The crack was large—an estimated twenty to twenty-three feet long (six to seven meters)—in a duct that was 197 feet long (60 meters). The chief fault lay in a heat-saving mechanism known as an economizer. The economizer is ordinarily meant to extract heat from the waste gases emitted during the combustion processes in the boiler and return as much heat as possible to the boiler.

One of the main reasons for the design fault was that the consortium of contractors had no experience designing a boiler of the size that Matimba required.[52] Corten-HT, otherwise known as "weathering steel," is a steel alloy designed to resist rust and erosion. It gained popularity in the United States during the 1930s, when it was used in railway wagons that transported coal, and it became commonly used in the transportation of heavy cargo. According to an engineer who worked for the Sieva consortium at the time of construction, the main problem was that the material had not been suitably stress relieved at the time of installation.[53]

During the subsequent stretches of litigation between Eskom and the consortium, the latter disputed the need for a complete overhaul, which would entail replacing the ducts. The contractor argued that the ducts required only regular repair work, which necessitated the complete stoppage of the power station's operations while the repairs were carried out. But Le Roux argued that the frequency of the required repairs would impede the power station's productivity and its ability to contribute electricity to the national grid: "So we disagreed," he said. "But they insisted that was the solution—it was a small change to the maintenance, and it was no problem.

Contested Neoliberalism ⁓ 83

Well the courts couldn't say whether it was or wasn't a problem. We ended up paying 50 percent of that repair."[54]

IMPROVED PRODUCTIVITY

In 1997 the National Productivity Institute awarded the Matimba power station the National Productivity Award. In his application for the award, Le Roux detailed the changes he had overseen at the power station to raise its productivity.[55] The power station's efficiency was demonstrated by the fact that it operated with relatively few disruptions to its activity. In 1996 it ran all six of the generating units for eighty days in a single stretch, beating the world record for uninterrupted operation.[56]

At times, the maintenance of core components meant that the power station had to be taken off-line and cease production for a short time. Because it generated such large quantities of heat, the cooling-down process was long and tedious. Part of Matimba's claim to improved efficiency in 1997 was that it had sped up its cooling-down and starting-up processes.[57] Under normal operating conditions, the boiler pipes channeled a continuous stream of hot steam. An innovative forced cooling process, which involved cooler steam being forced into the boiler pipes optimized production. Matimba's engineers boasted that they had successfully cooled down the power station in eighteen hours, a record time for all the power stations in Eskom's fleet.[58] Andile Williams,[59] an African novice power station engineer whom Le Roux had recruited as part of an inaugural staff overhaul, had accomplished this feat.

While a qualified engineer, Williams had no work experience at a power station, having been previously employed as an assembly line manager at a Lever Brothers toothpaste factory on the Johannesburg East Rand. As the only African engineer at Matimba, Williams found his colleagues to be less than willing mentors, and he encountered open hostility to his presence. After he experienced the same hostility in different sections of the power station, and in response to his eventual exasperation, Le Roux promoted him to the position of production manager of two generating units. Le Roux saw this as a calculated risk because Matimba's productivity had by then improved enough to allow him to relax his drive for technical excellence. Taking advantage of Williams's acute intelligence as well as his technical naivete, Le Roux set him on what was broadly considered an impossible technical feat. He describes it as follows:

> In that time, the average time to repair the boiler when one of
> the tubes burst was eighty-four hours.... So I said to Williams, "I

want you to find a way to repair the boiler in forty-two hours." Why? Because Williams didn't know that it couldn't be done. And I didn't believe my guys that it couldn't be done in less than twenty-four. But I didn't want Williams to do it in eighty-two; I needed a real step change, that's why I said half, in forty-two. Williams went out there, but it took a few months before he got his first tube leak on unit three. And he did the repair, and he did it in forty-seven or forty-eight hours. Now that was the record—it was the record by twenty hours. Fastest tube leak repair ever. [This was] because he spent the next two months asking stupid questions: "Why does it take you thirty-five hours to cool the boiler?" And they said, "Because the air that we're pushing through the boiler is hot, and it takes a long time to cool a boiler down before we can go in to inspect where the tube is broken first." He said, "Isn't there a way to cool it down quicker?" And they'd say, "No, that's just the way we do it." Conservativism. And the more and more he asked questions, the more they were not giving him logical answers. And because he was a very clever man, he found the illogic in their answers. So he questioned them further and further and further. And he found a way to do it quicker, to speed up different parts of the process.[60]

Eventually, Williams found a way to repair the boiler in eighteen hours, and the new technique spread throughout Eskom's fleet of power stations to become engrained operational wisdom. In this way, Le Roux advocated what he described as a "liberal" view, which he contrasted with the more "conservative" approach of some of his contemporaries. Technological innovation, he argued, was at odds with a centralized command approach, which he termed a "disempowering process," praising instead the importance of individual autonomy in solving mechanical problems.[61] This demonstrates the immersive character of the neoliberal transition in the management reform of the power station, including an ethos that cherished individual worker autonomy.

RACIAL TRANSFORMATION

Fitting with the apartheid government's technopolitical project after 1976, Eskom had made tentative moves to electrify the Black townships, particularly those on the outskirts of Johannesburg.[62] Ian McRae, the CEO of Eskom from 1985 to 1994, describes in his memoirs a clandestine meeting he held with ANC members in Soweto. He had first contacted John Rees and Bishop Peter Storey at the Central Methodist Church in Johannesburg, and the two

Contested Neoliberalism ⁓ 85

accompanied him to the Regina Mundi Church in Soweto, which was a long-standing site of community activism in the struggle against apartheid. McRae writes that apart from his driver, Moses Metsweni, no one at Eskom was aware of this meeting. McRae was struck by the underdevelopment of the Black townships in the Johannesburg periphery and the worrying implications of material inequality for a successful democratic transition.[63]

Le Roux had some experience with the project of racial integration, having previously been the manager of the Majuba power station in the Eastern Transvaal, where he started in 1987. The Majuba power station was built during the 1980s as part of Eskom's power station six-pack, but its completion coincided with Eskom's dawning realization that it had too much electricity generation capacity, and Majuba was mothballed once construction was complete. Even though the power station was inoperative, a contingent Eskom staff worked at the power station and lived in its vicinity. This made it an appealing site for an experiment in racial integration because inoperativeness reduced the bargaining power of the White trade unions. Work stoppages were to no avail since there was no pressure on the power station to produce electricity, and protests against the loosening of racial segregation in residential areas and in the workplace fell on deaf ears. The town nearest to the Majuba power station is called Volksrust, and Le Roux was tasked with introducing racial integration there, which meant loosening racial housing restrictions for Eskom's staff in White residential neighborhoods. He was assisted by an oft-ignored clause in the Group Areas legislation at the time, which exempted Eskom from the act, so the residential districts for Eskom's staff could thus legally be racially mixed.[64]

In addition to improving Matimba's finances, Le Roux also carried a mandate from Megawatt Park to encourage racial integration and employment equity at the power station starting in 1992. When Le Roux arrived in Ellisras in 1992, the NP and ANC were negotiating the impending political transition and the government was in the process of scrapping the Group Areas Act. Le Roux found it helpful to solicit the support of the Ellisras police station commander. He had neglected to win over the police station commander at Volksrust during his tenure at the Majuba power station, and the property of the few Black employees who ventured to inhabit houses in the White residential areas had been violently sabotaged. Le Roux felt that the support of the police chief in Ellisras would prevent official legitimation of vigilante activity. He convinced the station commander of the legality of mixed-race settlements and alleviated his fears that the impending transition would give rise to an attack on White townsfolk.

After a review of Matimba's staff contingent within the first few months of his arrival, Le Roux fired nine out of the ten managers who reported directly to him and began a rehiring process to replace them. Then, turning his attention lower down the hierarchy, he dismissed seventy out of one hundred of the first line managers.[65] To observers such as Steve Kekana, who was the worker representative of the National Union of Metalworkers in South Africa at the power station, Le Roux made a clean sweep of it: "The role that he played here was that he removed all heads of section, heads of function, heads of department. He removed them all."[66]

As part of the bid to encourage racial integration, Le Roux organized a weekend retreat for the mostly White power station workers where TV personality Freek Robinson joined as a special guest. Robinson was known as the face of the current affairs TV show *Fokus,* then broadcast on the pay television channel M-Net, and Le Roux hoped Robinson's influence could drive home the reality of racial transformation for workers. Racial integration was at times coercive, as Kekana related: "On a bus with three seats on one side and two on the other, you'd find a black wouldn't sit with a white and there'd be empty spaces. When Clive [Le Roux] saw this, he said the bus must stop and then put us one, one, one, one. He said he didn't want to see two blacks sit together or two whites sit together."[67] In this way, the valorization of the free market combined with an imperative of racial integration that would inform the nature of Matimba's transition into the democratic era.

South African government officials and Eskom's managers partially incorporated tenets associated with the neoliberal orthodoxy during the late 1980s and early 1990s in relation to the reform of the South African state corporations. The initial wave of privatization began in the late 1980s as the fiscal crisis of the apartheid government became apparent and featured individuals such as Wim de Villiers and Clive Le Roux, who worked for the government and Eskom respectively. These individuals believed in the efficiency of the free market and sought to guard against the excesses of central planners. But in the end, Eskom escaped privatization because both the outgoing NP and the incoming ANC government saw the value of its remaining tethered to the government as an agent of technopolitical rule.

At the Matimba power station, the project of neoliberal reform involved a break with past conservativism, valorizing instead the autonomy of the

Contested Neoliberalism ⁓ 87

power station engineers as a means of achieving technical innovation and saving costs. The cost-saving imperative also gave rise to court action with contractors, opening up a degree of technical uncertainty that was only resolved through the compromise of both parties. During this period of profound change, Eskom attempted to undo the prevailing practices of racial segregation at the Matimba power station, urging transformation in the notoriously conservative district of Ellisras. Eskom's network was thus able to adopt to the political and economic imperatives of its time and ensure its continued survival despite the splintering impetus characteristic of the neoliberal period.

5 ∽ Labor and Belonging in Lephalale

THE STORY of labor in the Waterberg, much like everywhere else in the country, is the story of the transformation from transience to permanence, or from a labor force classified as migrant to one with rights of belonging to cities and towns. For most of the twentieth century, corporations in South Africa relied on a deeply embedded system of low-wage migrant labor.[1] Within this labor regime, forged in the entrails of the gold mines, male mineworkers endured spartan living conditions in the mine compounds and spent their paltry wages on the maintenance of rural homesteads where their families ordinarily resided. As the twentieth century wore on and apartheid gathered up more and more Africans to live on the 13 percent of the country's land reserved for the homelands, it became increasingly difficult for homeland residents to eke out a living. By the 1980s, hunger and overcrowding characterized the homelands, and the migrant labor system as it had existed was no longer tenable.

The question of the permanency of an urban African population was a recurring one in South Africa's political life in the preceding decades, and its actualization was characterized more by negotiation and compromise than by the unfolding of a master plan. The apartheid government, once in power, had refused to allow African urbanization to run a natural course.[2]

89

In keeping with the principles of racial segregation while ensuring a steady supply of labor to the cities, the government established townships on the outskirts of major cities, overseen by the Group Areas Board, while preserving the urban core for Whites. Africans gained the rights to the city in a piecemeal way and this struggle, much like in the period leading up to decolonization across the African continent, was driven by trade unions and other platforms of worker solidarity.[3] Worker organization in the town of Ellisras, the name of which changed to Lephalale in 2002, followed the coal mine—power station networks that had developed elsewhere in the country, particularly in the Vaal Triangle in the south of the Gauteng province. Trade unions helped workers to negotiate the transition from a labor regime based on paternalism and dependency to one in which the free market nominally instituted individual autonomy. Timothy Mitchell describes the socio-technical organization in the coal mines in certain European countries as having set the conditions for workers to make democratic claims on the state. In Lephalale, worker organization also cemented the rights of African workers to the benefits of belonging in a context where they had unceremoniously been made strangers in their own land.

The bargaining power of African workers improved in the 1970s, and their growing assertiveness led to demands for the improvement of their living conditions. The Riekert Commission of 1978 drew government policy toward "stabilization," or the recognition of an urban African workforce. It also entailed the transition from a system of dependency to one in which individual freedom nominally reigned supreme.

But in Lephalale, at least, this came up against a history of capital dispossession, and trade unions emerged to negotiate the transition.[4] While the history of African trade unions in South Africa has largely been written as a history of the city, dynamics in the countryside were no less pivotal in the country's labor history. Due to the obsessively extensive reach of the Group Areas Board, Africans in the countryside were denied residential rights in small towns across the country. Border industries, where the government incentivized corporations to situate their factories at the borders of the homelands, proved largely transitory. Even during their brief existence, the various plants faced little challenge from unionized workers because of the desperate economic conditions of the homelands from which they sourced labor. But for Lephalale, which held far more permanent state industrial activity because of the state corporations, trade unions emerged to negotiate the masses of poor and dispossessed into a world where racial barriers were nominally removed.

Underlying the operations of the Matimba power station and the Grootegeluk coal mine was a long history of under-skilling of Africans in Lephalale. In accordance with the South African color bar, White workers occupied highly skilled positions and enjoyed trade union membership. African workers, whose preserve was manual labor, were barred from joining trade unions until the 1970s. While labor relations were decidedly brutal, the system was sustained by bureaucratic coercion that rested on a convoluted body of labor legislation, rather than on a consistent show of force.[5] The arrival of independent African trade unions represented a fundamental shift in the bargaining position of African workers, and trade unions assisted with the navigation to a system in which workers were nominally economically free. This led to the weary realization of the perils of neoliberal freedom among power station management and trade union officials alike, some years before the end of apartheid. This chapter reveals, on the one hand, the difficulties inherent to the neoliberal transition in the region, and on the other, the way in which Eskom ensured the continuing resilience of its network by adapting to the changing political prerogatives of democratic participation.

LABOR RELATIONS DURING THE TWENTIETH CENTURY

South Africa reaped the benefits of cheap coal until the 1970s; this was due to relatively cheap migrant labor both from within and without the country, with workers from all across southern Africa traipsing to the country's various mines.[6] But decolonization in Mozambique in 1975 and signs of hostility from authorities in Malawi and Lesotho threatened to disrupt these labor-sending channels to South Africa.[7] South African corporations reduced their reliance on migrant labor from neighboring southern African countries and encouraged the "South Africanization"[8] of the African workforce, which had the effect of strengthening the bargaining power of South African workers.

Wages paid to African workers had started to rise in the mid-1960s, mostly in the manufacturing sector, and this encouraged the increasingly confident African trade union movement of the 1970s. As a result, the wages paid to African workers steadily increased. While they continued to earn far less than their White counterparts, the gap between the two began to narrow.[9] In response to the situation of relatively scarce and expensive African labor, corporations invested in machines to reduce their demand for unskilled labor. This turn to mechanization also led to a shortage of skilled labor, and corporations in the private sector began to call, in ever greater

Labor and Belonging in Lephalale ～ 91

numbers, for the relaxation of the color bar. While Africans were placed in skilled positions, the racial integration of the workplace was only successful, as Feinstein writes, in small industries where belligerent White trade unions were relatively scarce.[10] The combination of a better-paid African workforce and the increased demand for skilled labor naturally led to the calls for "stabilization." The 1979 Riekert Commission, tasked with investigating the implications for influx control of the changed labor regime, urged the recognition of a permanent African skilled and semiskilled workforce in urban areas.[11]

BUREAUCRATIC COERCION

The migrant labor system that predominated for much of the twentieth century was not entirely based on brutality and involved some measure of worker consensus at the place of work.[12] In his description of the "moral economy" governing relations between managers and workers in the gold mines during the 1940s, Dunbar Moodie writes: "The mobilization of the entire workforce in moral outrage was an infrequent and hazardous exercise. But it did happen, and the fact that it might happen placed definite limits on complete autocracy."[13] The term *paternalism* implies that management viewed African workers as childlike dependents, incapable of autonomous decision-making. While paternalist labor management allowed workers a degree of leeway, the absence of bureaucratic norms meant that its impact was variable and dependent on personal relations between workers and management.[14] African workers at the Matimba power station were only able to enter into negotiations with management on an equal footing after the formalization of trade union engagement during the 1990s.

Despite its remote position in the country's northwest border, Lephalale was a heavily securitized pocket of the bushveld because of the apartheid government's fear of guerrilla incursion from neighboring African countries. A nearby military base bolstered the power of the town's police station commander. Because of the high security risk, the Matimba power station was considered a national key point, which meant that it enjoyed special military protection, a designation it holds to this day. When construction at Matimba began, reporters for the *Rand Daily Mail* described the intensive security measures at the power station site, which included "9.5 kilometres [6 miles] of double, barbed wire-topped security fencing around Matimba's looha [hectare] site [247 acres], with patrol roads on the perimeter and para-military guards at all access points."[15] According to Stephen Kekana, who worked at the power station during the 1980s, the security guards at the

92 ⁓ *Apartheid's Leviathan*

power station were a force to be reckoned with: "The security [at Matimba] then trained like soldiers," he said.[16] Labor strikes were entirely illegal and almost immediately curtailed if they began.

A plethora of labor laws underpinned this system so that it operated through a form of bureaucratic coercion. In 1979, the Riekert Commission remarked on the unnecessary complexity of the labor legislation, being "particularly struck by the extensive, complicated, and in many respects, fragmentary and overlapping measures, i.e. statutes, regulations, administrative rules and practices having a bearing on manpower matters in South Africa."[17] The Department of Labor ensured that corporations complied with the legislation by conducting routine inspections. During Iscor's initial prospecting work in Lephalale in the early 1970s, a labor inspector visited at regular intervals to investigate the accommodation and discipline of African workers. The inspector was preoccupied with reducing labor turbulence and the consequent loss in productivity, which meant ensuring that the management of labor contained an optimal combination of coercive control and worker acquiescence.

When Iscor began its prospecting work for suitable sites to situate the coal mine in the region during the late 1960s, many of the African workers were redeployed from Iscor's iron ore mine in the nearby town of Thabazimbi, which had opened in the 1930s.[18] The prospecting work necessitated constant mobility as the workers moved from site to site, and they lived in makeshift housing, generally in huts built of zinc. Iscor had to acquire the Department of Labor's approval for these houses. In its initial application, it requested clarity on the amount of "air space" each worker was allowed by government regulation.[19] In response, the director of Bantu Labor wrote that each hut ought to house six workers to allow for "200 cubic feet of air space per person."[20]

In June 1971, a labor inspector visiting a hostel that housed Iscor's African workers described the labor conditions he encountered. He found that, though the temporary hostel had been approved in April 1971, it had not yet fulfilled the requirements, and he ordered the demolition of the zinc huts housing three families. Despite the fact that the huts had been recently constructed at the time of the inspector's visit, he thought it probable that they were already "smothered with germs."[21] The inspector was, however, most bothered by a strike that was in process at the time of his visit by workers who sought higher wages.

Following another visit in March 1972, the inspector reported that conditions at the workers' accommodation had vastly improved. He praised the

Labor and Belonging in Lephalale ⁓ 93

efforts of the new hostel manager who had properly fenced off the living quarters of African workers, increased the number of policemen on site and improved the quality of the food. A general "improvement of living conditions" meant that African workers had become acquiescent and that appropriate measures were taken to prevent future strike action. The inspector also suggested that the placid labor situation was assisted by the presence of migrant laborers from Botswana who sought employment at the coal mine at the time.[22]

THE SPIRITS OF LEPHALALE

When Iscor's workers initially excavated the site for the Grootegeluk coal mine, they uncovered the bones of people who had not received proper burials and were likely to have perished alone. This legacy formed the founding mythology of the African township Marapong, the name of which translates to "place of bones." A resident of the Marapong township and traditional healer, Lazarus Seodisa, belonged to the prospecting team that initially traversed the Lephalale district in search of a suitable site for a coal mine before he found employment at Grootegeluk during its early excavation. He was there when workers found the sets of human bones. Community members identified one of these, by the cloth found in its immediate surrounds, as a woman called Sara Moloantoa. She hailed from the nearby village of Seleka and had gone missing after leaving her family on 24 December 1953 to visit her brother who lived on a nearby farm for Christmas. Travelers on foot would have faced a vast expanse of dry bushveld and deep darkness after sunset, and it was thought that she had lost her way and perished.[23]

According to Seodisa, workers uncovered other bones at the site they thought were most likely to have been the remains of people from the neighboring countries of Botswana and Zimbabwe. These travelers would have crossed the border into South Africa with the hopes of finding work at the iron ore mine of Thabazimbi, or heading further south to the gold mines on the Johannesburg Rand. In addition, Seodisa believes that fighters from the African National Congress's armed wing, Umkhonto we Sizwe (MK), who died in the Lephalale district were buried at the current site of the Medupi power station. More recently, in 2015, a group from Marapong led by Seodisa was trying to gain an admission from the managers of Medupi that they had come across the bones of human bodies during the excavation of the site. These bodies were thought to belong to MK fighters who had died and been buried on the site.[24]

In the early years of Iscor's prospecting, Seodisa thought that the angry spirits sabotaged the establishment of coal mines at various sites. Workers considered digging shafts in a site close to the Seleka village and another close to the dumping grounds in Onverwacht. At these sites, water flooded out of the first holes that they dug, and Seodisa interpreted this as a sign of the spirits' unhappiness at the rude disturbance. At the site of the Grootegeluk coal mine, excavated during the early 1970s, workers insisted that the bones be exhumed and relocated to nearby gravesites, and Iscor's management watched as traditional healers performed the necessary cleansing ceremonies.[25] A proper burial was considered necessary to lay the spirits of the dead to rest, and without it they would wreak havoc for the living. Seodisa argued that the spirits had to be approached with reverence and caution if the mine was to successfully sink a shaft.

The unrest of the spirits has become an important interpretive theme for Seodisa in his narrative of events in the region. As part of his work as a traditional healer in the community, he lobbied the Department of Traditional Affairs to consider the relocation of graves at the site of the Medupi power station. At Medupi, he argued, the spirits of the bodies who had not been laid to rest caused supernatural incidents and freak accidents for the workers at the power station, particularly those on the night shift. He argued that workers at Medupi who were born and bred in Marapong were generally aware of the role of spiritual forces in the region.[26] Eskom eventually exhumed these graves at the Medupi site and reburied the remains in accordance with the appropriate burial rites.[27]

WORKER SOLIDARITY

Once the Grootegeluk coal mine was fully operational in the mid-1970s, the balance of power between management and labor was variable and heavily dependent on contingencies. Seodisa described having assumed the role of an informal mediator between management and workers during periods of labor strife. He realized the value of playing off tensions between an Australian manager on site and the Afrikaner managers, calling on the former's assistance when African workers were mistreated by the Afrikaner workers on site.[28]

Nevertheless, this paternalist labor regime rested on a hierarchy of power that was structurally intolerant of dissent, and African workers could not always rely on the manipulation of interpersonal relationships. Another early employee of the coal mine and current resident of Marapong, Hendrick Ndebele, recalls an instance in 1981 when African workers at Grootegeluk

Labor and Belonging in Lephalale ～ 95

went on strike to protest the single-male hostel accommodation, demanding that housing for married couples be provided. Workers wanted their families to live with them and be able to seek employment in the town of Lephalale. Tight security at the hostel meant that there was little opportunity for workers to reside with their families informally, under the official radar.[29] The hostel was also situated in a distant location, necessitating daily transport from the hostel to the mine. Ndebele recalls being made to stand upright in the back of the truck during this commute. Despite worker discontent, buses with seating places were introduced only during the 1990s.[30] Thus, there were limits to the concessions worker unrest could achieve, and reforms were enacted at the behest of management.

Melton Mothoni, who worked at the Matimba power station during its construction period and later became the National Union of Mineworkers' main representative at Matimba, described the absence of formal negotiation structures for African workers within the power station during the 1980s. As a result, management was not obliged to seriously consider the demands of the African workforce.[31] Stephen Kekana, who joined Matimba's employ in 1985 and introduced the Metal and Allied Workers Union to the power station, described African worker activism as directionless before the arrival of the trade unions. Workers, he argued, were like a "flock of sheep."[32]

There were, however, clear instances of strike action during this period. They were either ended by a show of force or taken seriously if management deigned to consider workers' demands. Stephen Kekana detailed his struggle to organize workers during the 1980s at the Matimba power station at a time when strikes were headed off by the security at the power station and the police force. When Kekana first began to work at Matimba in 1985, there were a few trade unions that organized workers at the power station. One of these, the Electricity Workers Union (EWU), was an internal union sponsored by Eskom, but workers easily lost faith in its organizational structures because they perceived them to be too closely tied to management. Another was called the Boiler Makers Union. While it had a multiracial membership, its leadership was White and African workers quickly left its fold once an alternative emerged. Employees that sought to unionize workers could submit stop order forms to management for distribution to workers, which allowed the subscription to be paid from workers' salaries to the incipient union. But there was a perception that management discarded the stop order forms if they disliked the idea of the proposed union. "We didn't have structures," Kekana says. "It was difficult to organize workers those years because we were not protected."[33]

96 ～ *Apartheid's Leviathan*

The legislation governing strike action and the mobilization of African workers was complex and, for budding trade unionists, open to manipulation. Kekana recounts his strategic manipulation of labor laws—revealing that when challenged, the complex body of bureaucracy governing the activity of African workers tripped itself up. When workers were indicted for illegal strike action, they exploited obvious contradictions in their defense. For instance, since Africans lacked citizenship rights outside of the homeland borders and were considered foreign workers in the White districts of industrial activity, they could not be subject to the prevailing regulations on strikes. As Kekana relates: "They told us we were in an illegal strike . . . but we told them no—those laws don't bind us because we did not have the right to work. That's how we used to argue." In this way, the complicated tangle of labor bureaucracy ran to the end of its tether when faced with the assertion of rights from the African workers it intended to keep in check.

TRADE UNIONS

African trade unions had emerged in a disparate manner during the 1970s, with many struggling to gain recognition from management and legitimacy in the eyes of the workers. One of these trade unions was the Metal and Allied Workers Union (MAWU), which was incorporated into the National Union of Metal Workers of South Africa (Numsa) in 1987. Kally Forrest writes that by the end of the 1970s, MAWU was on the "brink of collapse, as membership fell, and by 1977 all organized workplaces had folded."[34] This was partly due to the fact that many companies had refused the union space to organize the workers in their employ. MAWU's waning popularity and consequent subscription crisis during this period was worsened by the brutal police shut down of a strike at the Heinemann Electrical Company in the Johannesburg East Rand, one of its most important sites of worker organization.

In 1979, South Africa's Commission of Enquiry on Labor Legislation, commonly known as the Wiehahn Commission, recommended that African workers be given the right to join trade unions, and this paved the way for employers to recognize African trade unions. Collective bargaining practices for African workers, which corporate managers accepted as a measure to alleviate industrial conflict, was a decided break from the past.[35] The power of the various trade unions grew apace and they demonstrated their full force when they cohered under the banner of the Congress of South African Trade Unions in July 1985.

Labor and Belonging in Lephalale ～ 97

Despite the difficulties it encountered in the 1970s, MAWU managed to improve its organizational strength during the 1980s and adopted strategies to maneuver around the restrictive labor legislation. To deal with the intransigence of individual factory managers who would not recognize the union, for instance, MAWU worked on entrenching the concessions to labor that had already been awarded in civil contracts.[36] The union's management also emphasized developing the shop floor strength of each factory to encourage an in-depth familiarity with its particular working conditions. As Forrest notes, budding trade unionists attended information sessions run by academics at the University of the Witwatersrand. These academics faulted preapartheid African trade unions such as the Industrial and Commercial Workers Union (led by Clements Kadalie) for organizing as "general trade unions." A sector-wide organizing strategy also alleviated the fears of individual factory managers that labor concessions would give their competitors the upper hand, as MAWU could point to similar worker demands at the factories of their competitors.[37]

The National Union of Mineworkers (NUM) and Numsa began organizing workers in Lephalale in the late 1980s, and they have remained the main trade unions in the district's coal mine and power station nexus. By organizing workers at the Matimba power station, both unions strayed from their traditional industrial sectors. NUM had traditionally organized mineworkers, while Numsa concentrated on the metals and engineering sector. The pattern of union organization followed the coal mine and power station nexus that existed elsewhere in the country.

Kekana argues that the involvement of both unions at Eskom's power stations is rooted in their prior experience in the industrial hub of the Vaal Triangle to the south of Johannesburg. MAWU had first begun organizing workers at an Iscor-owned steel factor in Vanderbijl Park. Two of Eskom's power stations, the Vaal and Lethabo power stations, are situated in the vicinity of Vanderbijl Park and drew their coal supply from the nearby coal mines where NUM organized mineworkers. MAWU enjoyed a strong regional presence because it mobilized workers at Iscor's steel factory, which encouraged the union to venture into the energy sector and organize workers at the nearby power station. In a similar manner, NUM first organized workers at the coal mines and then signed up workers at Eskom's nearby power stations. In this way, both NUM and Numsa (which incorporated MAWU in 1987) gained experience with organizing power station workers, which eased their organizational access to the Matimba power station in Lephalale.

98 ⁓ *Apartheid's Leviathan*

NUMSA AND THE MATIMBA POWER STATION

The arrival of MAWU at the Matimba power station represented the first attempts at independent and concerted African worker organization at the power station. Kekana played a formative role in introducing the union to the power station workers. Before the arrival of MAWU, he had worked as a shop steward for the EWU. Following a workshop for EWU shop stewards in Lephalale, Kekana hitched a ride to the nearby town of Nylstroom with one of the instructors who was on his way back to Germiston, on the Johannesburg East Rand. On his return, the instructor took the idea of recruiting workers at Matimba to MAWU's head office, and the union deployed two of its members to the power station to help Kekana recruit members and gain recognition from the power station management.

After MAWU morphed into Numsa in 1987, the union built up its strength at the power station against the backdrop of the repressive state of emergency that began in 1985. In 1987 Kekana successfully defended a worker who faced disciplinary action after a coal-carrying conveyor caught fire on his watch. Fires on coal conveyors are a common hazard, caused when the friction between the rollers and the carrying surface of the conveyor generates enough heat to ignite the coal being carried on the conveyor belt. The worker had not paid enough attention to the process to prevent the occurrence, and Kekana argued that his neglect was due to an illness that the worker had reported to his supervisor beforehand. This victory encouraged other workers to place their faith in Numsa's organizing structures, opening the door to further recruitment. According to Kekana, workers from different departments at the power station subsequently joined the union's fold in large numbers.[38]

NEGOTIATIONS AT MATIMBA

Once African trade unions achieved formal recognition at the power station, they were absorbed into a formal negotiation process that oversaw the power station's transition into the democratic era. As the Matimba power station manager starting in 1992, Clive Le Roux introduced regular meetings that allowed for consultation between power station management and representatives of the union MWU-Solidarity, which historically had a White membership, Numsa, and NUM. One of the most pressing agenda items was the need for the different parties to reach consensus on the way that affirmative action would play out at the power station, especially as it related to recruitment and training policies. The minutes of the meetings in which the unions and management hammered out the details of transformation

Labor and Belonging in Lephalale ⁓ 99

reflect a surprising degree of consensus from the trade unions that transcended racial divisions. This concurrence rested on their desire to protect the rights of their members already in the power station's employ and to ensure fair consideration of these workers for staff appointments.

The minutes of a meeting of the affirmative action task group held in August 1994 provide useful insight into the viewpoints of the various trade unions.[39] MWU-Solidarity disagreed with the idea of "token" appointments from the outset. The notion of token appointments here refers to the practice of hiring employees on the basis of their skin color to fill affirmative action quotas while neglecting their actual competency. While not explicitly rejecting the policy of affirmative action, MWU-Solidarity's representative argued that management should first consider internal candidates, or those candidates already in the employ of the power station. In this formulation, a suitably qualified White internal candidate should be preferentially appointed over an affirmative action candidate drawn from outside the ranks of both the power station and Eskom. Numsa's representative concurred with MWU-Solidarity on the need to avoid tokenism but disagreed on the order of preference for appointments. Numsa agreed that White internal candidates be considered before an affirmative action candidate from outside of Matimba and Eskom. However, the union argued that the needs of affirmative action candidates at Matimba and then Eskom were paramount, and were to be considered before internal White candidates. NUM attempted to refine the criteria for appointing internal affirmative action candidates, stating that "minimum qualifications" were to be applied in considering these candidates, while "maximum qualifications" were relevant for White employees.[40]

For the next few months, the recruiting relationship appeared to run smoothly, and in September 1994 the unions "expressed their appreciation of the way Engineering consults with them in terms of recruitment and transparency thereof."[41] However, the trade unions continually propagated the importance of internal candidates in their views on subsequent appointments. This demand often came up against the shortage of competent internal candidates. In June 1995, an employee in the electrical maintenance department proposed the appointment of an affirmative action candidate for a position in welding in the boiler.[42] While the appointment was eventually approved at the meeting, NUM's representatives voiced their concern that the proposed candidate was not an internal one. Rather, "they felt that Matimbans irrelevant of race should get first preference." But management responded that no internal candidates had applied for the position and that those potential affirmative action candidates in the "accelerated development program" at Matimba were not yet suitably equipped to perform the job satisfactorily.[43] In this way, the imperatives of stabilization led to concerted efforts for local residents to enjoy the opportunities for employment and skills development that the power station offered.

Certain senior managers at Matimba appreciated the value of employing internal candidates, though out of necessity rather than virtue. At a meeting held on 7 August 1995, the head of Matimba's engineering department spoke of his travels to the Eastern Cape town of East London to investigate the possibility of appointing affirmative action candidates from the region.[44] East London was an industrial hub because of the large amount of motor manufacturing plants situated there. But he found on arrival that none of the twenty-five candidates he interviewed met the technical requirements for employment at Matimba, demonstrating the limits to the recruitment of external candidates. He thought it more fruitful to focus on training employees within the power station and proposed to fill three vacant positions with trainees from Matimba. NUM and Numsa supported the proposal and added that the danger was that appointments from outside of Matimba (and possibly Lephalale) would simply use the power station as a "stepping stone" before disappearing.[45]

One of the problems with hiring internal candidates was that the training pipeline could not match the power station's demand fast enough. In 1994 Le Roux complained of a shortage of engineers at a national level so that, even with the power station's engineers then in training, there was likely to be a shortfall by the year 2001. The situation worsened if the pool was narrowed to include only engineers who qualified as affirmative action candidates because they were not graduating from universities in the vast quantities required.[46] The shortage of skilled engineers and artisans at the time constituted a national skills crisis because of the historic exclusion of Black students from institutions of higher learning.

However, some workers had become technically competent by learning on the job, and there emerged a retrospective process of assessment, then being implemented across the country, known as the Recognition of Prior Learning (RPL).[47] Proponents of RPL argued that workers should be eligible for promotion on the basis of their experiential learning, and the government put in place measures to determine an employee's actual efficacy and ascertain whether or not they fulfilled the necessary occupational requirements.[48] Kekana relates that certain employees at Matimba were successfully promoted on the basis of RPL, such as plant operators who

Labor and Belonging in Lephalale ⁓ 101

had learned how to handle certain equipment on the job.[49] Staff members performing administrative tasks also achieved certification through RPL processes. Given that there were no tertiary education institutions in Lephalale, engineers with university degrees came from outside of the district. RPL enabled the promotion of internal candidates, generally from the Lephalale district, who were not able to access tertiary education. NUM asserted that by "watching someone do a job one should be able to do it" and that "'spanner boys' after 10 or 15 years could be artisans."[50] They also argued that theoretical training was tied to competency in the English language, and this posed a language barrier for those workers who were unaccustomed to the technical terminology.[51] NUM continued to insist that transformation occurred too slowly at the power station, demonstrated by what they considered to be the sluggish upward movement of workers through the employment hierarchy. Complaints persist to the present day of the partial and long-drawn-out nature of the certification process for RPL candidates.[52]

Relations between management and trade unions broke down in March 1997, when the trade unions declared a dispute and embarked on a weeklong strike. The matter necessitated the intervention of Eskom's executive director of generation at the time, Bruce Crookes, who aimed to restore the trust between management and labor. The seeming persistence of racial segregation animated the protest. Numsa demanded the removal of racially segregated toilets, which were a legacy of Matimba's construction during the latter years of apartheid.[53] An official bulletin released by the power station management stated that the "implementation of affirmative action" was one of the chief concerns, and the neglect of "people from Matimba" was particularly at issue in training positions and presumably also in opportunities for promotion.[54]

In August 1997, the unions and management reached some consensus about the way the power station would handle Crooke's recommendations. On the policy of recruitment, the eventual decision was that Matimba would revert to Eskom's power station–wide policy on affirmative action, rather than rely on its own Matimba-specific one.[55] The main implication of this transformed policy was that all positions would only be advertised externally when all possible means of recruiting from within the power station were exhausted. Trade unions were to "influence the compilation of the job description and the profile of the desired incumbent."[56] But this rigorous consultation at the early stages of recruitment allowed management some autonomy when deciding on the final candidate. A memorandum detailing the new recruitment and selection procedure stated that the "final selection of successful candidates is now left to the discretion of line management

102 ⁓ *Apartheid's Leviathan*

and needs not to be reported back at a sub BU forum."[57] While it is difficult to gauge the success of these policies in satisfying the different constituencies, it is clear that there was little certainty from the outset about the way that transformation was to proceed and that it was negotiated into a permanent state of irresolution.

THE PERILS OF THE FREE MARKET

At Matimba, the perils of the free market are ably illustrated in the issue of salary deductions, a practice that Matimba's managers attempted to end after 1992. But NUM resisted because it considered payroll deductions to be a safeguard for workers who would sometimes neglect to make mundane but essential payments if left to their own devices. One of these was to a company called Istores, although the nature of the services it provided is not clear. In May 1998, NUM urged management to reinstate the system of stop orders from employees' payrolls and proposed that Istores impose limits on the amount that workers could purchase, noting that "some individuals did not manage their money."[58] A similar debate arose over the question of the salary deductions for payments to a pension fund company called Iemas NUM protested the proposed cancellation of the payments, and a member of the Matimba management team stated his concern that workers were having difficulty managing their money and were "using Eskom as a tool to control finances."[59] Then, in May 2000, unions urged management to continue deducting water and lights payments from workers' payrolls, fearing that if the practice was stopped, workers would not make the payments themselves and their water and electricity would be disconnected.[60]

In 1998 the power station began to phase out the hostel-based residential system, signaling the transition from a single-male system based on coercion and surveillance to one where residential life was subject to the wiles of the free market. While the hostel was in existence, Matimba's management had to rule on the question of allowing female visitors to enter the hostel premises, and because there were multiple workers to a room, they opted to ban women from the hostels. Le Roux described it as follows:

> The apartheid system created dependency. People were dependent. The system told you what you could and couldn't do, so you didn't have choice. And we were the system. We had to make those decisions: can you bring a woman into your room or not—and you're fifty-five years old or thirty-five even, you know, more independent. So those were the things we were dealing with—all of us,

Labor and Belonging in Lephalale ⁓ 103

not just at Matimba. All the power stations, the whole of Eskom was dealing with those dynamics in the 1990s.[61]

The decision to finally close the hostel was a product of both ideology and necessity. The hostel was a massive structure, built to house five thousand laborers during the time of construction.[62] By the mid-1990s, workers occupied only a tiny portion and the rest of the hostel fell into disrepair. Vandals steadily stripped the near vacant hostel of those fixtures which still held some market value. Electric cables were popular items of theft, but removing them was a dangerous exercise and at times the bodies of those who had tried were discovered on the hostel premises.

An alternative to hostel accommodation for African workers was the single quarters, which housed Eskom's White, skilled workforce, who were generally early-career single males. The single quarters accommodated one person per room, which contained a bed, desk, and closet. But the accommodation at the single quarters was more expensive than in the hostel, and Le Roux recalled that, while hostel dwellers paid R5 ($0.27) per month for accommodation and meals at hostels, residents at the single quarters paid R105 ($5.72).[63] As a result, some workers refused to leave the hostels. Workers who moved to the single quarters complained that the meals there contained more vegetables and less starch and meat than the food traditionally served at the hostels.[64] But NUM and Numsa encouraged the closure of the hostel—in line with the unions' national campaign—because of its visceral connection to the brutality of the colonial and apartheid regimes.

Over time, the number of residents in the hostel gradually declined so that, in July 1998, Matimba's management reported that only fourteen employees remained.[65] According to Kekana, the intransigent workers who remained at the hostel were migrant laborers from the KwaZulu-Natal and the Eastern Cape provinces. He suggested that with a permanent homestead in these regions, the migrant laborer was more willing to live a transient life at his place of work.[66] With so few inhabitants, the hostel became financially unsustainable, setting the stage for its eventual closure. But the mass migration from the hostel caused some consternation among those employees already resident in the single quarters. One particular complaint was of the pockets of prostitution that followed the hostel dwellers to the single quarters.[67] Thus the hostel closure, which symbolically signaled a shift away from the migrant labor system, was met with unexpected resistance from some workers.

Across Eskom's fleet of power stations, managers introduced market-related rentals to Eskom-owned houses. The Eskom Housing Policy of 1995 stated that, in line with the Reconstruction and Development Program and national housing policy, Eskom committed to find housing for its employees and, in particular, to assist employees toward owning their own homes.[68] Demand for houses at the lower end of Eskom's rental grade exceeded supply, and African employees were entitled to a settling-in allowance from Eskom, which allowed them to move to married quarters. This was a first for many because the dictates of the Group Areas Act had outlawed their residence there in previous years.[69]

～

Before South Africa realized the enfranchisement of voters of all races in the 1994 elections, African trade unions functioned as important sites of democratic participation in the workplace. This chapter has highlighted the role of democratic participation, in this case through trade union bargaining forums, in contesting the imposition of policies in line with neoliberal orthodoxy at the Matimba power station. This contestation was, in turn, a product of the legacy of dispossession in the region, which saw Africans forcibly removed from land in the vicinity of the town of Lephalale to the nearby homeland. The transition from a paternal, autocratic system of labor management to one in which the free market reigned supreme was thus a complicated one. The management at Matimba introduced systems to address the demands of transformation and autochthony, and these remained contested processes. These included, for instance, addressing the historical exclusion of the country's Black majority from institutions of higher education as well as balancing the needs for racial transformation with that of localization.

Labor and Belonging in Lephalale ～ 105

6 ~ The Medupi Power Station

WHEN THE African National Congress (ANC) came into power in 1994, it promised not only to realize political liberation, but also to deliver the fruits of modernity to the previously disenfranchised majority of the country. Infrastructure and technology took center stage in this mission. In 2007, South African first began to experience the blight of load shedding, or scheduled periods of electricity outages. Eskom would announce the onset of load shedding for fixed periods of time, when technical faults or scheduled maintenance at its power stations meant that it would not be able to satisfy the country's electricity demand. In response to the underlying electricity shortage, Eskom announced its first new construction projects of the postapartheid era: the Medupi and Kusile power stations. Medupi would be built in the Lephalale district, close to the Matimba power station, and Kusile would be constructed in the Mpumalanga province. The stakes were undoubtedly high for the completion of Medupi and Kusile because their electricity would ensure that the country's economic growth would not falter for want of power. But the completion of the power stations took much longer than expected—despite Eskom's initial estimation that they would be completed by the end of 2013. The delay in the completion of Medupi and Kusile thus became a source of much public angst as load shedding continued to afflict South African households and businesses.

As with other major infrastructure construction projects of the 2010s, Medupi has been subject to accusations of corruption. While political interference certainly played a part, it was not the sole reason for the delay in the power station's completion, especially when one considers the fact that Medupi's construction outlasted the successive presidencies of Thabo Mbeki and Jacob Zuma. Medupi's construction continued into the presidency of Cyril Ramaphosa, who took office in 2018 and promised to crack down on systemic corruption in the structures of the state. Despite the best political will to complete the power station construction, Medupi succumbed to technical failures that proved costly and time-consuming to rectify.

The Commission of Inquiry into Allegations of State Capture, set up in 2018 by President Zuma to investigate allegations of state corruption (although most of its hearings took place under the presidency of Cyril Ramaphosa), spent a significant amount of time on events that occurred at Eskom during the 2010s. The term "state capture" came to be used in South Africa to describe the usurpation of ministerial powers, notably by the notorious Gupta family, headed by three brothers, who left the country in 2018 for their numerous properties in Dubai. The Gupta family arrived in South Africa in 1993 as small-time businessmen and made inroads into the ANC after gaining the confidence of Thabo Mbeki in the mid-2000s. But it was only under the presidency of Jacob Zuma that they reached the top levers of the government and accessed the power to divert public funds into their offshore bank accounts. They have to date not repaid any part of the estimated $3 billion they had siphoned over two decades.[1] During the hearings of the state capture commission, it emerged that Medupi had been drawn into the dealings of the state capture operatives. These operatives had taken action against certain executive members of Eskom whom they blamed for Medupi's delay in 2015. Though they had no basis for these accusations, it served as a pretext to appoint their own subordinates to executive positions at Eskom and ultimately aid in the project of diverting funds from Eskom to private individuals.

The construction of Eskom's new power stations coincided with a period of fiscal expansion in South Africa. In response to the recession wrought by the global market crash of 2009, the governments of Brazil, Russia, India, and South Africa, collectively known as BRICS, adopted counter-cyclical policies aimed at stimulating the economy through government spending.[2] Around the same time, emerging markets—of which South Africa was one—became an attractive investment destination because of their relatively high bond yields. The inflow of capital played an important role in keeping South Africa's balance of payments account steady. In addition, the World Bank encouraged African countries to invest in new infrastructure, particularly in their energy generation capacity. In 2010 the World Bank launched its Africa Infrastructure Diagnostic Report, which lamented Africa's scarce infrastructure and the prospects for economic growth on the continent.[3] In subsequent years, construction began on large infrastructure projects across the continent, including the Grand Ethiopian Renaissance Dam, the DRC's Inga 3 Dam, Kenya's standard gauge railway from Nairobi to Mombasa, and Nigeria's Mambilla hydroelectric power station.

South Africa was best placed to begin a program of government-driven economic stimulus in 2009 because a decade of fiscal austerity had created relatively sound finances, and national debt was 30 percent of GDP. This period in turn coincided with the political rule of President Zuma, whose reign has come to be characterized by a spending exuberance. Encouraging the state corporations to invest in new projects, the government guaranteed the debt not only of Eskom but also of Transnet, the state corporation responsible for the railways (previously called South African Railways and Harbours).

The cost overruns of the Medupi and Kusile have made up a large proportion of Eskom's total debt. In 2009 Eskom embarked on what it termed a "Capital Expenditure Program," in which it envisioned spending R385 billion ($21 billion) over five years. To raise money, it issued bonds in the form of a domestic multiterm note program, which was listed on the Johannesburg Stock Exchange. The government agreed to stand as guarantor for a maximum of R175 billion ($10 billion) of Eskom's debt, and this was formalized in an agreement signed in November 2009 by the minister of finance at the time, Pravin Gordhan.[4] In 2012 Eskom expanded the size of its Capital Expenditure Program and increased the amount to R453 billion ($25 billion). In response, the minister of public enterprises, Malusi Gigaba, acting with the "concurrence" of the minister of finance, increased the amount it would guarantee to R350 billion ($19 billion).[5] Through information revealed at the hearings of the state capture commission, Gigaba has been closely linked to the Gupta family.[6]

Officials working at Transnet have also been implicated in receiving kickbacks for the award of contracts for locomotives and telecoms services, among other things.[7] The Passenger Rail Agency of South Africa, which was once a part of Transnet but separated from it in 2009, has also seen an excessive wastage of funds. Its former CEO, Lucky Montana, oversaw the procurement of R5.3 billion ($290 million) worth of Spanish manufactured locomotives that proved too tall for the country's rail network when they arrived in the country in 2015 and were auctioned off in 2019.[8] By December

108 ~ *Apartheid's Leviathan*

2019, the country's debt had reached 62.5 percent of GDP and ratings agencies endowed the country with a junk credit status, so diminishing its ability to raise money on international credit markets.

While probably the largest individual profiteer, the Guptas were not the only recipients of the spoils of corruption. Corruption had seeped into many levels of government.[9] The flouting of official rules on procurement blurred the line between party funding and personal enrichment so that civil servants at times created personal fiefdoms.[10] In December 2017, Cyril Ramaphosa was elected president of the ANC at the party's elective conference in a nail-biting finish, defeating Zuma's preferred successor, Nkosazana Dlamini-Zuma. Zuma lost the legal aid of the state and was cast out to the judicial wilderness, appearing in court on multiple occasions on allegations of corruption on his own personal account. But this political change came too late to stem the deterioration of public finances. The country's GDP growth declined over the next three years, exacerbated by the periodic return of load shedding.

During the 2010s, South Africa benefited from a rise in mineral exports. China was a rising world power, with a large appetite for raw materials, particularly iron ore from South Africa, so boosting the fortunes of South African mining corporations. Then, in December 2019, Eskom reached "stage six" load shedding, which was the highest level it had ever reached in terms of the regional spread and duration of the controlled blackouts. Gold mines in the Carletonville region on the outskirts of Southern Johannesburg had to suspend their operations to prevent mineworkers from being stuck underground in the case of electricity outages. In March 2020, AngloGold, which had operated in South Africa for over a century under the control of the Oppenheimer dynasty, sold off its last remaining gold mine in the country, which is also the deepest mine in the world. In so doing, it collapsed the tacit pact between Eskom and the gold mines that had driven the South African economy for the entirety of the twentieth century.[11]

THE CONSTRUCTION OF MEDUPI

Medupi was hastily designed while South Africa was in the thick of an electricity shortage. Eskom's laggardness in developing new generation capacity was due to its having been cowed in the 1990s by the pressures of austerity. President Thabo Mbeki was fixated on neoliberal austerity until 2004, when the government switched gears and announced the start of a new era of development, elucidated in the country's national development plan. In the early 2000s, Eskom warned that South Africa faced an electricity supply shortage, but the government did not take heed until 2007, and President Mbeki then publicly apologized to Eskom and admitted to the seriousness of the electricity supply problem.[12]

In a press interview in September 2013, Brian Dames, Eskom's CEO at the time that Medupi was commissioned, said that its construction might have run smoother if Eskom had started planning earlier, but because of the haste required, Eskom used virtual designs based on its Majuba power station, which was also dry-cooled.[13] Majuba and Kendal were also dry-cooled power stations and the immediate predecessors of Medupi and Kusile. But Eskom found itself in a weak purchasing position toward the end of the first decade of the twenty-first century because of high global demand for power station equipment. Buoyed by the rising industrial powers of India and China, the handful of global multinational corporations that manufactured power station equipment were swamped with orders.

At the time of its inception, Medupi's design was high-tech and entirely novel. In order to cope with the arid climate of the region, Medupi is a dry-cooled power station, so that it uses less water than the more conventional wet-cooled power plants. In order to limit carbon dioxide emissions and alleviate environmental concerns, the boiler is of a supercritical type, which means that it has a relatively high pressure to assist in heat generation while burning less coal than a conventional boiler. Medupi is designed to generate 4,800 MW of electricity, making it the fourth largest coal-fired power station in the world.

At the time of Medupi's design, nowhere in the world was a dry-cooled power station with a supercritical boiler and a generation capacity of that size in existence. Part of the reason for the technological novelty of Medupi was that Eskom had bucked the global trend of renewable energy, which was the main area of innovation in energy generation. The haste with which Medupi and Kusile were planned allowed Eskom to bypass objections about the environmental impact of coal-fired power stations. Eskom generally viewed environmental organizations with suspicion, and in 2013 it was forced to apologize to Greenpeace for spying on its activities. Its large institutional funders, the World Bank and the African Development Bank (AfDB), confronted the wrath of environmental organizations after they agreed to fund the power station.[14] Eskom's engineers believed that renewable energy technology was too unreliable at that point to satisfy the scale of South Africa's electricity demand. The World Bank had advocated infrastructural investment, particularly electricity development, across Africa, and it agreed to loan Eskom $3.75 billion in 2010.[15] The AfDB agreed to lend

110 ～ *Apartheid's Leviathan*

Eskom €1.86 billion (an estimated R21 billion at the time) at the end of 2009.[16] Both the World Bank and the AfDB argued that the construction of a coal-fueled power station was the only way in which South Africa's electricity demand forecast could be met, which would in turn ensure that the country reached its projected GDP growth of 4 percent.[17]

ANC, HITACHI, AND CHANCELLOR HOUSE

The award of the tender for the supercritical boilers at the new power stations proved to be a major controversy early on. Eskom awarded the tender, worth R38.5 billion ($2.1 billion), to Hitachi Power Africa, a South African subsidiary of the Japanese-based global multinational corporation Hitachi. This decision raised red flags because the ANC owned 25 percent of the company's shares through its investment arm, Chancellor House.

In 2009, the public protector launched an investigation into Eskom's tender decision-making process following a request from Helen Zille, the leader of South Africa's main opposition party, the Democratic Alliance. The public protector is an office independent of the government, headed by an advocate of the law, and set up under the aegis of the South African constitution, to investigate allegations of impropriety of public officials. The chairperson of Eskom, Valli Moosa, was also a member of the ANC's fundraising committee at the time that the contract was awarded, and his role in the award of the tender came under scrutiny during the investigation. The *Mail and Guardian* reported that the same month Eskom decided to construct Medupi, Chancellor House acquired a 25 percent stake in the shares of Hitachi Power Africa. Moosa had been the chairperson of Eskom for four months at the time.[18] In his evidence to the public protector's commission of inquiry, however, Moosa argued that Eskom's tender process was sound and that there was no room for an "arbitrary" ruling on his part. His assertion was backed by the findings of an independent review conducted by the auditing firm Deloitte, which concluded that the tender process had been fair. While the public protector, Advocate Lawrence Mushwana, concluded that the tender had not been improperly awarded, he faulted Moosa for his failure to manage his conflict of interests in accordance with Eskom's policy. This potential conflict of interest arose because of Moosa's dual position as a member of the ANC's National Executive Committee and as the chairperson of Eskom.[19]

The controversy died down following the public protector's investigation, and the matter remained a surprising coincidence until October 2015, when an investigation by the United States' Securities and Exchange

The Medupi Power Station ⌁ 111

Commission (SEC) revealed that Hitachi had "inaccurately recorded improper payments" to the ANC. The payments consisted of a $1 million "success fee" and $5 million in dividends from the ANC's shares in the company. These payments were related to the award of the Medupi boiler contract.[20] Hitachi chose not to defend itself and agreed instead to pay a $19 million fine to settle the charges.

The evidence unearthed by the SEC's investigation, which included internal company emails, together with Hitachi's unwillingness to launch a defense, offers compelling reasons to suspect a relationship of impropriety between the ANC and Hitachi. The SEC's investigation revealed that the relationship between Chancellor House and Hitachi began in late 2005. The following year, Hitachi submitted its tender application for the boilers and turbines at Medupi, and similar tenders for the Kusile power station in 2007. In May 2007, an executive member of Hitachi Power Africa (HPA) emailed their colleagues in the company, stating that while the company "'had not been successful in receiving any update [from Eskom],' Chancellor House and 'HPA's 5% shareholder' were doing 'their very best' to bring Hitachi's offer for the Eskom contracts 'in first place.'"[21] But in August 2007, Eskom awarded the boiler contract to the French multinational corporation, Alstom, which was Hitachi's only rival in the tender race. In an interview I conducted with a senior Eskom official in 2015, the official argued that if Hitachi and Chancellor House indeed worked hand in hand, their initial failure was proof of the fact that the latter did not have a direct influence on the tendering process.[22]

Nonetheless, Hitachi eventually acquired the contract following the breakdown of negotiations between Alstom and Eskom. The SEC argued that Hitachi had gotten wind of the fact that Eskom was having difficulties negotiating its terms with Alstom. Hitachi consequently "directed Chancellor House to help Hitachi win reconsideration of the boiler component of the Medupi power station contract."[23] By July 2008, Chancellor House had received its "success" fee together with the dividend payments from its shares. In addition, the SEC alleged that when Hitachi bought back its shares from Chancellor House to dampen the surrounding controversy, the latter received a 5,000 percent return on its shares. The evidence points to an improper relationship between Chancellor House and Hitachi, but the manner in which Chancellor House influenced Eskom's tender process is unclear.

Regardless of whether or not Eskom subverted its ordinary tendering process and awarded the boiler contract to Hitachi due to political favor, it is not clear that this action caused the technical problems that subsequently emerged in the boiler. Hitachi was not a new entrant to the South African electricity generation market, and it had a long relationship with Eskom through its ownership of the boiler manufacturing company Steinmüller. Steinmüller had manufactured and installed the boilers at the Matimba power station in the 1980s. The German corporation Babcock Borsig bought Steinmüller in 1999, and then in 2003, Hitachi bought the energy division of Babcock Borsig.[24]

Neither Eskom nor its subcontractors had any experience installing supercritical boilers, and they encountered a shortage of skilled artisans who could perform the required welding. According to interviewees who were employed at Hitachi at the time of Medupi's construction, the boiler at Medupi was fundamentally different from any boiler installed at any of Eskom's power stations, including its Waterberg predecessor Matimba. The supercritical technology utilized in the boiler meant that the boiler needed to withstand far higher temperatures and pressure than conventional subcritical boilers.[25]

Despite the boiler contractor's familiarity with the South African context, the installation of the boiler did not run smoothly. In March 2013, the *Business Day* reported that Eskom suspected the reports it received from Hitachi detailing progress on the boiler work had concealed the true nature of the welding faults. Eskom's investigations had revealed that some of the welding performed on the boiler at Hitachi's factory in the Johannesburg East Rand town of Nigel did not meet the required specifications and that "9000 pressure welds had not been post-weld heat-treated."[26] Paul O'Flaherty, Eskom's chief financial officer at the time, claimed that Eskom had repeatedly warned Hitachi about potential problems with the welding procedures, saying, "We have been telling Hitachi we don't believe you are conforming to your specifications and they have said, 'Don't worry, we'll take the risk, we'll take the risk.'"[27] The welding was a delicate operation, consisting of thousands of intricate welds to connect pipes and equipment. In September 2013, the Hitachi Power Africa representative responsible for both Medupi and Kusile, Mark Marais, discussed the consequence of the lack of competent welders in the country at a seminar hosted by the Department of Science and Technology. The power stations required a particular type of welding called mirror welding, which meant that the welder would have to perform the weld on the basis of a reflected image in a mirror before him. Commenting on the complexity of the operation, Marais quipped: "I can't even tie my shoelaces while looking in the mirror."[28]

The Medupi Power Station ⁓ 113

The factory in question was situated in the Johannesburg East Rand town of Nigel and was run by a subcontractor of Hitachi, DB Thermal. DB Thermal also had a long-standing presence in South Africa as a supplier of cooling equipment to the mines, dating back to the 1970s. In 2009, in partnership with Hitachi, it invested in the development of a new boiler manufacturing facility. Its stated motivation in expanding it factory in Nigel in 2009 was to fulfill Eskom's 60 percent local manufacturing requirement by establishing a local manufacturing presence and to invest in training a skilled welding workforce. This development was an extension of the company's already existing manufacturing facilities in Nigel, which were initially established during the 1970s to supply the boilers for Sasol's electricity generating units. The company was also involved in manufacturing parts for Eskom's new power stations in the 1980s.

On 25 July 2013, one of the subcontractors responsible for welding on the boiler was charged with fraud. Hitachi's forensic investigations discovered that the subcontractor had submitted fraudulent documents and covered up the extent of irregularity in the boiler work.[29] After ongoing investigations, Eskom reported in September 2013 that the nature of the fraud was so sophisticated that it had gone undetected by the professional bodies responsible for quality assurance. Eskom asserted that the subcontractor had used incorrect welding procedures in the boiler equipment it intended to supply to Medupi. The welding procedures did not fit Eskom's quality control assessments and thus could not confidently withstand the high heat and air pressures that would be created within the supercritical boiler. In addition, the postweld heat treatment was incorrectly performed.[30] Postweld treatment is applied to the material surrounding the weld, which is ordinarily damaged during the intense heat treatment of the welding process. A former Hitachi employee described the problem as due to the fact that "it was a new type of boiler tube material (which required higher welding skill) coupled with inexperience and aggravated by poor management and QA [Quality Assurance] processes."[31] The technical failure was thus due to a combination of inexperienced welders employed at the Nigel factory and the failure of quality assurance management to notice that the welding had not fulfilled the requirements.

The repair work involved a team of about fifty workers carrying out repairs on the welds, and they managed to repair the boilers so that they fit the requisite welding technical specifications.[32] In addition, a subcontractor had to replace four separators, which are devices used to separate water from steam within the boiler system.[33] Eventually, in November 2013, Hitachi

114 ~ *Apartheid's Leviathan*

reported that its boiler work at Medupi was mainly complete. This followed Eskom's postponement of Medupi's commission date from the end of 2013 to the second half of 2014.[34] Medupi's budget had also been revised upward, from R91.2 billion ($5.3 billion) to R105 billion ($6.1 billion), excluding interest. In September 2013, Eskom's commercial director, Dan Marokane, stated to the parliamentary portfolio committee that "we will be going after contractors,"[35] and Eskom's effort to recoup at least some of the costs led to legal battles that went all the way to the Supreme Court of Appeal (SCA). Eskom sought to utilize its right to call in its performance bonds against Hitachi as penalties for non-delivery. Performance bonds are a commonly used mechanism in capital-intensive projects, intended to function as a guarantee for the client if the contractor fails to deliver. The funds are set aside by the contractor as part of its contractual obligation. The SCA eventually ruled in Eskom's favor in September 2013, upholding Eskom's right to call in its performance bonds from the Mizhuo Corporate Bank of Japan, which had acted as guarantor for Hitachi's work for Eskom.[36] The guarantees amounted to about R600 million ($34.8 million).

STATE CAPTURE COMMISSION

Dan Marokane was appointed to the position of Eskom's acting group executive for group capital in 2013 after the resignation of Eskom's chief financial officer, Paul O'Flaherty, who had been heavily involved in the construction of Medupi. O'Flaherty's departure followed Eskom's announcement that the first unit of Medupi would not be complete by the end of that year, as was initially promised. Marokane fully assumed the position in November 2014 and oversaw the process of bringing the first power generating unit of Medupi online by the end of 2015, though he was abruptly suspended from Eskom in 2015. He had become caught up in the state capture saga, known to the public through explosive media reports that President Zuma had delegated his ministerial prerogatives to the wealthy Gupta family for their personal enrichment. In October 2020, Marokane testified at the state capture commission, and his testimony to the commission included chilling tales of officials being unceremoniously fired for no apparent misdeed and of a host of shadowy figures linked to the Gupta family, who had pulled the strings behind the scenes at Eskom and other state corporations. The video recordings and transcripts of the hearings of the state capture commission are available online and freely accessible to the public.[37] They provide valuable insight into events that occurred at Eskom during the height of the state capture period.

The Medupi Power Station 〜 115

In 2015 the board of Eskom, chaired by Zola Tsotsi, suspended four executive staff members, namely Tshediso Matona, Tsholofelo Molefe, Dan Marokane, and Matshela Koko—without charging any of them. These executive members, with the possible exception of Matshela Koko, were thoroughly mystified by their suspension. The board of Eskom told the executive members only that they were being suspended pending an investigation into their responsibility for the continuing electricity shortages and the delays in the completion of the Medupi and Kusile power stations. This action was hastened by a complaint laid by Minister of Public Enterprises Lynne Brown in the months leading up their suspension. In a media statement dated 12 March 2015, Brown stated that she had shared her concerns with the board of Eskom about issues broadly related to the country's ongoing electricity shortage and the persistence of load shedding. Brown's complaints specifically mentioned the "overruns at Medupi and Kusile" and the "progress with the build programme."[38] Brown has since been implicated in the dealings of the Gupta family through her personal assistant Kim Davids. In 2017, media reports alleged that the Guptas had paid for Davids's airfare to Dubai and for her chauffeured drive to the Gupta's house in the city.[39]

The irony of the suspension of at least two of the executive members, Marokane and Matona, is that they believed themselves to have helped remedy Eskom's problems, which began before they were appointed to their respective positions. In his testimony to the state capture commission, Marokane said that within a month of his appointment, he had identified and communicated to the board three main problem areas at Medupi that required resolution. These were the "boiler welding defects," the "control and instrumentation system non-compliance," and the "creation of a stable environment for a productive worker force."[40] Marokane then created problem-solving teams that consisted of representatives of Eskom and the contractors. He also claimed to have built on new labor agreements implemented by his predecessor, O'Flaherty, to identify disputes early on, and to have implemented a "new integrated schedule," which encouraged transparency because it integrated risks across the construction site and allowed the creation of a new cost estimate. Marokane said that he had "created meaningful and tangible progress in the execution of various projects which saw, in particular, the delivery of the first unit of Medupi—achieved one week prior to my suspension."[41] The findings of an investigation, conducted by law firm Dentons, eventually found no evidence of wrongdoing by any individual, though it detailed issues related to delays and cost overruns at Medupi in broad terms, such as the hasty design at the outset, inadequate quality

116 ⌁ *Apartheid's Leviathan*

control, and poor project integration. It did not in any way attribute the problems with the new power stations to the suspended executives, nor did it find that Medupi's major technical faults occurred during their tenure.

A little while before Marokane's suspension, the board had suspended Eskom CEO Tshediso Matona in March 2015. Matona told the state capture commission that in June or July of 2015, President Zuma had requested a private meeting with him at the presidential office. At the meeting, Zuma told Matona that his suspension held no bearing on his "person or professional character" and that he hoped Matona would continue to work in the public sector, lamenting the fact that Matona had gotten "caught in the middle of a spaghetti."[42]

Like Marokane, Matona was appointed in October 2014, well after the commissioning and construction of the new power stations. His own perception of the work he had accomplished was positive, and he highlighted the fact that his suspension (and that of the other executive members) led ratings agency Standard & Poor (S&P) to downgrade Eskom's long-term credit rating to junk on 19 March 2015. S&P attributed its decision to Eskom's suspension of Matona and the other executive members of Eskom. Matona and Marokane said it was clear from their engagement with Eskom board members that returning to their respective positions was not on the table, and they reached resignation settlement agreements.

The testimony to the state capture commission related above demonstrates that the operatives of state-level corruption used the seemingly technical nature of Medupi's delay to remove competent executives from Eskom for their own nefarious ends. Their justification involved the assertion that responsibility for the problems at Medupi could be attributed to a few individuals, when in reality the faults were complex and multifaceted. The various disparate technical and human elements combined in the construction of the power station became so entangled that they were ultimately inseparable.

LABOR

In addition to faults in the power station equipment, labor unrest and work stoppages added to Medupi's construction delays. These issues surrounding labor reflected the changing technopolitical configuration of the power station that was related to events and changes occurring in other parts of the country. At first, the consortium of contractor companies, as the entity with boots on the ground, acted as the chief project managers, and Eskom assumed a non-interventionist role in managing labor relations. But Eskom

The Medupi Power Station ⁓ 117

eventually intervened in the management of labor in an effort to encourage labor stability and reduce work stoppages. When President Zuma visited the Medupi construction site in 2012 for a pressure test on the first boiler unit, he announced Medupi's potential for massive job creation. But the bulk of the jobs created were low skilled and lasted only during the civil works construction phase. By 2015 workers were being laid off en masse, and the problem of chronic unemployment in the region remained.

The first labor agreement at Medupi, known as the Project Labor Agreement (PLA), was signed in December 2008 at the Mogol Club in Lephalale. It was signed by the contracting companies, which had formed a consortium called the Medupi Power Station Joint Venture (MPS-JV), and the local-level shop stewards of the national trade unions, chief among which were the National Union of Mineworkers (NUM) and the National Union of Metalworkers of South Africa (Numsa). While the agreement boasted the signature of labor representatives, the Numsa national organizer at the time, Steven Nhlapho, maintained that the agreement was not signed at the national bargaining forum and that the signatories were local shop stewards, who signed without being fully cognizant of the implications of the curtailment of labor rights contained within the PLA.[43]

The PLA proved to be a cause of conflict because it curbed workers' rights to protest in the interests of "minimizing risks" and promoting "labor stability." Controversially, it banned all strike action. The dispute resolution procedure laid out in the PLA stated that grievances had to be reported to the project industrial relations manager and escalated, if necessary, to the project dispute resolution committee for further arbitration. Neither party could refer the matter to the Commission for Conciliation, Mediation, and Arbitration, a national body for the resolution of labor disputes, or resort to any form of external arbitration. The internal channels of mediation fell under employer control, creating the perception of a compromised impartiality among union leaders. In addition the PLA did not specify how legal impartiality would be ensured.[44] It is not clear whether or not each employee had to be unionized, but the PLA specified that contractors could automatically deduct union fees from the wages of union members and later deliver the collected funds to union headquarters.[45] It also made it the responsibility of the contractor to provide furnished offices for shop stewards. At first, this took the form of interim facilities in Lephalale with the view to eventually relocate to an area closer to the construction site once land was available.[46] But the achievement of a modicum of consensus for the signatories of the PLA was not enough to head off labor unrest. Instead,

118 ⁓ *Apartheid's Leviathan*

discontent over the PLA led to the very work stoppages that drafters of the PLA sought to prevent.

In 2013, recurring and increasingly violent strike action paralyzed construction at the Medupi construction site. The first strike was reportedly initiated by eleven hundred Alstom employees who were installing the power station's controls and instrumentation system. They were soon joined by Hitachi's employees,[47] and on 16 January, work at the construction site came to a halt. Numsa's branch deputy secretary at the time, Seanego Ngakamone, told the *Star:* "We are not going to return to work unless the employer terminates the project labor agreement. They are not applying it in fairness, they are applying the part that favours them most."[48] He accused the employers of refusing to act in good faith. The strike continued with various instances of off-site vandalism. In mid-February, police arrested forty-six workers at the Medupi construction site for acts of violence that occurred in the Marapong township as a part of the protest action.[49]

Eskom was always present in the background, encouraging contractors and workers to resume work as quickly as possible, and it publicly disagreed with its contractors' employment practices. In July 2013, Paul O'Flaherty, Eskom's chief financial officer, said: "The ability of our contractors to supervise labor has been poor across the board. They have to adequately train the person to do the job and supervise them."[50] On 22 February 2013, the minister of public enterprises, Malusi Gigaba, intervened in the dispute.[51] A few weeks later, Gigaba announced that the strike had ended after his intervention, and he argued that harsher penalties needed to be imposed on the contractors for lost production time as an incentive for improved productivity.[52] While Gigaba's role in bringing the parties together is not entirely clear, the trade unions, contractors, and Eskom reached a new labor settlement in the immediate aftermath of his intervention.

In June 2013, the new agreement, known simply as the Partnership Agreement (PA), came into force. Among other things, the PA committed the signatories to greater wage standardization across the different contracting companies. The drafters of the PLA had considered the question of wage standardization but argued that it was impossible to completely standardize wages and working conditions across the site because of "individual contractor employment agreements." It had, however, prescribed a minimum wage for each job category. In addition, Eskom was a signatory to the new agreement, which meant that it assumed some of the responsibility for labor relations.[53]

Following the new agreement, strikes continued, although on a much more limited scale. On 25 July 2013, a strike occurred following a parliamentary

The Medupi Power Station ⁓ 119

oversight committee's visit to the construction site. Between five hundred and one thousand workers protested over the subject of allowances. Vandalism during the strike, which saw five vehicles set alight and several graders (a type of construction equipment) being pelted with stones, led to the arrest of forty-five workers.[54] The workers were charged with "malicious damage to property, public violence and unlawful gathering," and ordered to appear in a court in the Limpopo province. Then, in August 2013, another strike occurred with instances of vandalism of vehicles and construction equipment; security staff was injured and the construction site was shut down.[55] Contractors labeled this strike the "last straw," threatening mass dismissals with the view to begin a massive rehiring process.[56] In response to the strike, Eskom raised the minimum wage across the construction site to R25 ($1.30) per hour, above the industry minimum wage of R20.50 ($1.12) per hour.

In the ensuing years, worker unrest assumed an increasingly subversive character, occurring outside the sphere of formal labor negotiations. During strike action in 2015, the erosion of worker faith in formal channels of negotiation is evident in the increasingly anarchic character of the protests. Numsa began to lose control over the strike action that sporadically occurred, and there were numerous instances of anonymous vandalism and of arson committed by masked protesters. Following labor unrest in August 2015, a Numsa official explained: "It was almost impossible to identify Numsa members in a crowd. If we are sure they are Numsa members, our jobs would be a bit easier."[57] These activities mirrored developments in labor organization elsewhere in the country, tragically represented by the Marikana massacre of 2012. Marikana is a town in a platinum-rich region of the North West province, and in August 2012 mineworkers at the Lonmin platinum mine embarked on strike action. These workers had lost faith in their traditional representative unions of NUM and Numsa, and neither Lonmin's managers nor the trade unions could control the striking workers. On 16 August 2012, police opened fire on a crowd of workers, killing thirty-four.

By the beginning of 2016, it was clear that a proportion of the remaining workers at the power station had lost faith in the organizing structures of Numsa and NUM. A new union, called the Liberated Metalworkers of South Africa (Limusa), made its presence felt at the Medupi power station. Limusa had been formed in July 2014 at the national level by the former secretary general of Numsa, Cedric Gina. The Congress of South African Trade Unions had expelled Numsa in 2014, and Limusa was intended to fill the empty space within COSATU left by Numsa's departure. Limusa thus organized

120 ~ *Apartheid's Leviathan*

workers in the same sector as Numsa. In March 2015, a former NUM shop steward had signed up some of the remaining workers at Medupi to Limusa's fold.

In August of 2021, Eskom announced that Medupi was finally complete. This was welcome news to a South African public in the midst of the COVID-19 pandemic, during which load shedding had periodically occurred. A week later, an explosion occurred at one of Medupi's six units, which meant that the unit's installations would have to be entirely replaced before it could become operational—a process that would likely take two additional years. Despite suspicions of sabotage, investigations revealed that the explosion was in fact caused by a lack of adherence to protocols during a procedure to determine the cause of a leak. Medupi's plight seemed never-ending. Meanwhile, by December 2021, just four out of six planned units of the Kusile power station (in the Mpumalanga province) were fully operational. Kusile had initially been planned for completion in 2014. This chapter has demonstrated the changing regimes, encompassing labor and technical aspects, in a construction project that has taken more than ten years to complete. While it is tempting to ascribe Medupi's failures to corruption that sacrificed technical efficiency, it is difficult to make the case that political interference itself caused the technical faults that the power station experienced. This made Medupi an unknowable entity, plagued by various pressures that included technical difficulties, party-level corruption, development in a destitute region, and Eskom's unyielding faith in its ability to build a high-tech power station, the likes of which had not been seen anywhere in the word. As revealed in the state capture commission, those who sought to profit off corrupt practices at Eskom instrumentalized this uncertainty to serve their interests.

The construction of Medupi coincided with a more general embrace of new infrastructure projects in the late 2000s, signifying the resurgence of government spending to stimulate economic growth. This was a break with the fiscal austerity that had characterized the first decade of postapartheid South Africa. But this manifestation of government spending on new infrastructure differed from its characterization in the 1960s. In the town of Lephalale and its immediate surrounds, democracy has endowed political rights to a population previously rendered non-citizens. Labor unrest has been a major cause of Medupi's delay. Despite Eskom's best efforts, attempts at centralizing the management of labor have escaped the control of

The Medupi Power Station 121

Eskom, contractors, and the trade unions that earlier played an important role in the transition to democracy. The murky involvement of Chancellor House in the award of the boiler contract for the power station is also a direct result of the ANC's need to raise funds for the maintenance of its party structures. This highlights the capacious role that Medupi has come to play by absorbing the many different imperatives of the democratic era into its decades-long construction timespan. But by the late 2010s, the finances of the government and of Eskom could no longer cope with the escalating costs brought about by the construction delay and the repair processes. Eskom's liquidity crisis in early 2018, which highlighted the precarity of its debt situation, signaled that it was dangling close to the edge of financial crisis of national proportions.

Conclusion

AFTER THE initial public shock at the scale of Eskom's indebtedness in 2018, its fortunes improved slightly, but its electricity generation infrastructure continued to deteriorate. By 2022, there appeared to be no end in sight to load shedding. While Medupi has almost comically defied completion, its near-completion in August 2021 came at a time when the destructive effects of climate change around the world were plain to see. As opposed to the period of its initial design, there had also been rapid advances in renewable energy technology, which reduced the technological advantage of fossil-fueled power stations. The ambivalence of Eskom's network of power stations was again apparent—on the one hand as a salve to the electricity shortage crisis, holding the promise to economic growth in a country with a very high unemployment rate, and on the other hand as a contributor to the ever-worsening climate change crisis.

The ambivalence of Eskom's position in South Africa stretches back to its founding in the 1920s, a time when the South African economy revolved around the fortunes of the gold mines. As the dominant economic activity of the first half of the twentieth century, gold mining shaped the migrant labor regime. Since mineworkers traveled from rural areas across southern Africa, the gold mines had far-reaching impacts on socioeconomic developments in the region. The state corporations, Iscor and Eskom, were drawn into the orbit of the gold mines, nominally to provide the auxiliary supply of electricity and steel. But they also served a distinctive developmental purpose, focused on the South African government's mission of improving the lot of "poor Whites" who resided in both urban and rural areas across South Africa and whose destitution threatened the project of white supremacy.

While the state corporations initially struggled to gain monopoly control over their respective sectors, they grew from strength to strength after the Second World War. Then, in the 1960s, they began to enjoy a close relationship with the apartheid government during a period of apartheid rule that closely fits the outlines of James Scott's notion of authoritarian high modernism. They were thus a part of the project of government planning within which science and technology loomed large on the path to modernization. The government also became systematically repressive as it realized the principles of racial segregation enshrined in apartheid. But rather than viewing the state corporations, their engineers, and technology as enjoying a seamless accord with the apartheid government, their relationship is viewed in this book as both autonomous and immersive.

Iscor and Eskom slotted into the apartheid government's imperative of industrial self-sufficiency during the 1960s, which would allow the country to survive economically despite the likely imposition of international sanctions. Iscor proceeded tenaciously to expand its steel production capacity in order to meet the demands of South African industry. But the climate of straitened financing in the 1970s, due chiefly to the oil crisis of 1973, strained the relationship between Iscor and the government. In the end, the scarcity of international credit combined with the intrusions of the global export market for coking coal led Iscor to the Waterberg—a corner of the country with coal reserves that it would not otherwise have exploited.

While signaling the beginning of the neoliberal order, the early 1970s also threw up a bundle of contradictions. Both Iscor and Eskom were in the midst of expansion plans determined by their earlier demand forecasts. These forecasts predicted increases in demand for steel and electricity and, for Iscor, meant that it would have to ensure the security of its supply of coking coal while global demand for South Africa's coal reserves increased. This combination ultimately enabled the exploitation of the Waterberg coalfields.

The infrastructure developments in the Waterberg over the last few decades, revolving around Eskom's power stations and Iscor's coal mine, have proven Janus-faced because they look backward to an authoritarian past and forward to the hope of the realization of democratic ideals. In describing

124 ～ *Apartheid's Leviathan*

the essential ambivalence behind the activities of the state corporations in South Africa, this book has argued that they were not simply tools in the arsenal of the apartheid government. The initial exploitation of the Waterberg coalfields illustrates this well because the region was of little strategic objective to the apartheid government. Instead, it was drawn into the orbit of the national industrialization project through the tenacious endeavors of Eskom's and Iscor's engineers to realize their expansion plans. While the underlying demand forecasts were indirectly tied to the urgency of realizing the nation-building project, they were nonetheless abstractions that drove infrastructural development in the country for two decades.

This book has also described the technologically mediated manner in which a peripheral region, such as the Waterberg, was incorporated into the ambit of governmental authority and control. For the apartheid government the modern city was one that was suitably segregated. The Group Areas Board was largely successful in meting out forced removals in all urban centers across the country, ensuring that each city and small town contained distinct, demarcated regions for the occupation of a single racial group. Iscor's activities in the Waterberg drew the government's attention to the disorderly way in which the town had developed, and the transmission of governmental power to the town was realized fully in the forced removals of Africans living in the vicinity of the town in the mid-1970s. The transmission of governmental power that underlay the modernism characteristic of this period was not a direct, linear process, consisting of command from the government and its adherence at the periphery, but was instead made up of a complex assemblage of contingent factors and interested parties, of which Iscor's infrastructure and engineers were a part. In this way, the modernism that enabled a suitably racially segregated town assumed an unpredictable form at the outset. When one considers the procession of events in hindsight, it appears as one filled with unforeseen elements, complicating the idea of the realization of a distinctive authoritarian rationality.

Though the grand apartheid project proceeded on its messy path, there remained an overarching planning framework and a belief in its abstraction, which enabled a cooperation of sorts between the government and Iscor's engineers. But in the 1980s, the economy took a turn for the worse and the townships were engulfed in near perpetual unrest, which challenged the apartheid government's ambition to contain people and things within their specially ordained roles and regions. In addition, the confidence with which Eskom had embarked on its expansion plan was also threatened as it faced a situation of a mismatched demand forecasting. The 1980s signaled the incoherence of the high modernism of the apartheid regime—one in which the

Conclusion ⌒ 125

conviction in rectitude and in purpose that characterized the previous two decades was called into question.

During this period of disorientation and disillusionment with long-term planning, neoliberalism held increasing appeal for apartheid reformers. The government eventually opted for Eskom's commercial reform rather than outright privatization, and while certain Eskom officials saw the value of the free market and of the privatization of Eskom, it remained a state corporation. The onset of democracy and the extension of the universal franchise meant that the incoming ANC-led government would be judged on its delivery of material benefits and electrification in a context of huge wealth inequality. In the postapartheid period, Eskom's status as a state-owned corporation was beneficial to keeping electricity tariffs affordable to the majority of the citizenry. In the particular case study presented in this book, trade unions negotiated the transition from a paternal system of labor management to one in which the free market supposedly reigned supreme. NUM and Numsa had become familiar with the networks of coal and steel in their organizing work in the Vaal Triangle, and they organized in a similar fashion in the Waterberg. But individual economic autonomy was not easily achieved, and the transition also entailed the pressure of autochthony and encouraging belonging in a place where Africans had unceremoniously been turned into strangers a few decades before.

The democratization of the 1990s meant that neoliberalism became a consistently contested process. As a state corporation that owns a national network of enormous power stations, Eskom has also existed alongside smaller, "micro" technologies, such as the prepaid meter for electricity, that more closely align with neoliberal precepts of discrete, individual technologies. As the ANC's reign wore on, government corruption became more prevalent and evident. Eskom's continued status as a state corporation might have been due to the fact that it was the best mechanism to ensure that electricity tariffs were affordable, but as the democratic era progressed, it became subject to the pressures of party funding.

The breakdown of much of South Africa's infrastructure in recent years, with electricity being the most conspicuous failure, has highlighted the importance of infrastructure and technology to the measure of peoples' satisfaction with the ANC-led government. Despite the best will in the world, the new Kusile and Medupi power stations, initially hailed as the panacea to the electricity shortage, failed to reach timely completion. As the years following the initial panic about Eskom's imminent financial ruin in 2018 brought no immediate end in sight, Eskom became a touchstone for measuring the health of modern-day politics. It became at once crucial to the

126 ⁓ *Apartheid's Leviathan*

country's political fortunes and too complex to reveal a simple diagnosis that could lead to repair. A large infrastructure project like Medupi has been able to absorb different pressures while its construction lumbered on and its costs escalated. As a result, Eskom was unable to add enough new generation capacity to the grid, and load shedding continued into the 2020s. In 2021, following a week of load shedding due to electricity supply constraints, Eskom announced that the Tutuka power station had been subject to sabotage. While Eskom did not announce the causes of the sabotage, these allegations coincided with a period of intense divisiveness within the ANC. These divisions manifested in violent uprisings in July 2021, following the arrest of former president Jacob Zuma. Electricity outages offer an avenue for manipulation because public frustration at electricity outages was easily channeled toward outrage at Ramaphosa's leadership.

The scholarship on democracy in African countries generally focuses on the fragility of these democratic systems and the ever-present specter of a return to an authoritarian past. These democratic systems are constantly under threat by the prevalence of coups d'état on the continent and by governments' steady encroachment on civil liberties, such as the freedom of the press and the independence of the judiciary. The tale told in this book reveals that infrastructure and its sustainability can pose an equally dangerous threat to the well-being of a relatively new democracy such as South Africa. The failure of the ANC-led government to deliver an undisrupted supply of electricity has led to major dissatisfaction with its rule that transcends race and class divisions. Democratic rights are not only exercised during elections and expressed in the choices of political parties but, as the persistence of service delivery protests demonstrates, in the technological apparatus through which citizens measure their quality of life and sustain their hopes for the future.

Yet, at the same time, the ANC-led government lacks the ability to control the process of infrastructure development—in this case the construction of new power stations—and guarantee successful completion. This is in part because the government is not a unified entity, but is riven by intraparty conflict, and decisions are subject to dispute within the ministerial cabinet. This has important implications for our understanding of African politics and democracy on the continent because it demonstrates the perpetual intertwining of the material or technological and challenges a definition of "politics" that is chiefly human centered. In so doing, it reveals the vitality of technology as playing a determinant role in the trajectory of historical events. The "state" is shown to be made up of a multiplicity of departments and individual personalities that can chafe against each other and

Conclusion ⁓ 127

that are tied to the local level through various material and organizational intermediaries.

In South Africa, the construction of new infrastructure is particularly fraught because of the prevalence of corruption, which has drawn the state corporations in its wake. The amassing of wealth is also the route to politicking within the ANC, since it allows prospective presidential candidates to launch their campaigns at national and provincial levels. The various state corporations responsible for infrastructure development, including but not restricted to Eskom, have proven to be easy targets for pilfering because they are manipulable by the ruling party. But at the same time, the ANC's involvement is not entirely to blame for the obduracy of infrastructural projects such as the Medupi and Kusile power stations. Eskom has escaped being fully shredded by the wiles of corrupt politicians, instead retaining a distinctive autonomy. This meant that during the state capture saga, politicians who sought to profit off tenders for infrastructure could not simply override Eskom's tender processes. For example, the notorious Gupta family, working in concert with Zuma, had to strategically deploy operatives to strategic positions within Eskom to ensure that they could turn contracts in their favor. This speaks to the importance of the technological nature of infrastructure projects, which is one reason for their inability to be wholly controlled by politicians.

At the end of 2021, South Africa secured $8.5 billion in grants and loans from the United States, Britain, France, and Germany to fund the transition away from coal.[1] Gwede Mantashe, the minister of minerals and energy, protested the elimination of coal-fueled electricity, threatening court action in order to continue with new coal-fired power stations that the ministry had planned.[2] A former trade union leader, Mantashe argued that the coal mines provided an important source of employment in a country with an excessively high unemployment rate. He also argued that it would be possible to develop "clean coal," which involved developing power stations' capacity to limit harmful emissions. But in October 2021, the Center for Research on Energy and Clean Air named Eskom the world's worst emitter of sulfur dioxide. In addition, Eskom was among the world's highest carbon dioxide polluters, despite its unfulfilled promise to install flue gas desulfurization technology at the Medupi power station.

Mantashe's statements are illustrative of the conflicting views surrounding the transition away from coal in South Africa. While coal mines provide employment in rural areas, they also damage the surface environment and diminish the quality of life for communities who live in their vicinity. In addition, the destruction of flora and fauna has led to protests

128 ⁓ *Apartheid's Leviathan*

from environmental conservationists of the wisdom of establishing new coal mines. In such a situation of conflicting imperatives, it is challenging to identify a neutral arbiter who is able to chart a course through these competing views. The ANC-led government cannot be considered a cohesive entity when it comes to energy policy. In the particular case of the transition to renewable energy, President Cyril Ramaphosa has gratefully accepted the funding offer of $8.5 billion, but Mantashe, a cabinet minister, has voiced open dissent to the idea of a total transition away from coal. In a context where it is necessary for the views of ANC party, officials, and government ministers to align, there is unlikely to be a strong consensus around the need to adopt renewable energy in the near future. This further highlights the complexity of the managing ambivalent infrastructures and materials amid the ever-worsening climate crisis.

Following Eskom's financial crisis in 2018, it went through a number of leadership changes in quick succession. Then in November 2019, the minister of public enterprises, Pravin Gordhan, announced the appointment of a new CEO, Andre de Ruyter, who had worked at Sasol for twenty years before serving as the CEO of the packaging company Nampak. In 2021 de Ruyter announced plans to unbundle parts of Eskom into the discrete units of transmission, generation, and distribution. While these new units would still be wholly owned by Eskom, unbundling took the corporation further down the road toward privatization than had been the case in decades, promising to break up the monolithic entity Eskom had become.[3] Nonetheless, this raised the ire of the National Union of Mineworkers, who protested that unbundling indicated an enthusiasm for privatization, which in turn threatened to increase the price of electricity.[4]

The story of the state corporations' activities in the Waterberg ultimately reveals their interweaving relationship with both the apartheid and the democratic governments. They were both a part of and separate from the sphere of governmental politics, and South Africa's political development is inextricably tied to the activities of the state corporations and the infrastructure they created. While the opening of the Waterberg coalfields in the 1970s and 1980s was largely inadvertent, and outside of the initial scope of the government's and the state corporations' long-term plans, it ultimately served an important strategic purpose through the attempt to remedy the electricity shortage in postapartheid South Africa. Eskom in particular, as a monolithic entity for most of the twentieth century, adapted to the imperatives of neoliberalism and democratization in the early 1990s. In so doing, it ensured the continuation of an energy regime based on fossil fuels, so contributing to the ever-worsening climate crisis.

Conclusion ⌒ 129

Notes

INTRODUCTION

1. "Koko's Fight to Stay at Eskom Resumes," *Fin24,* 16 February 2018, https://www.fin24.com/Economy/Eskom/kokos-fight-to-stay-at-eskom-resumes-20180216-2.
2. "Eskom Too Big to Fail and Won't Be Privatized, Ramaphosa Tell Investors," *Fin24,* 15 May 2019, https://www.fin24.com/Economy/eskom-too-big-to-fail-and-wont-be-privatised-ramaphosa-tells-invetsors-20190515.
3. Here and elsewhere in the book, sums in rands have been converted to dollar estimates using rand-dollar exchange rates of October and November 2022.
4. Stuart Jones, *Banking and Business in South Africa* (New York: St. Martin's, 1988), 21.
5. Brian Larkin, "The Politics and Poetics of Infrastructure," *Annual Review of Anthropology* 42, no. 1 (2013): 327–43.
6. Steve Woolgar, ed., *Virtual Society? Technology, Cyberbole, Reality* (Oxford: Oxford University Press, 2002); Steve Woolgar and Geoff Cooper, "Do Artefacts Have Ambivalence? Moses' Bridges, Winner's Bridges and Other Urban Legends in S&TS," *Social Studies of Science* 29, no. 3 (June 1999): 433–49. Brian Larkin has also remarked on the changing meaning of infrastructures, but from the perspective of the user experience of already existing infrastructure: Larkin, "The Politics and Poetics of Infrastructure."
7. Frederick Cooper, *Africa since 1940: The Past of the Present,* New Approaches to African History (Cambridge: Cambridge University Press, 2002).
8. James C. Scott, *Seeing Like a State: How Certain Schemes to Improve the Human Condition Have Failed* (New Haven, CT: Yale University Press, 1998), 87–89.
9. Faeeza Ballim, "The Un-Making of the Group Areas Act: Local Resistance and Commercial Power in the Small Town of Mokopane," *South African Historical Journal* 69, no. 4 (2 October 2017): 568–82; Laurine Platzky and Cherryl Walker, *The Surplus People: Forced Removals in South Africa* (Johannesburg: Ravan, 1985).

10. Stephan Miescher, "Building the City of the Future: Visions and Experiences of Modernity in Ghana's Akosombo Township," *Journal of African History* 53, no. 3 (November 2012): 367–90.

11. Renfrew Christie, *Electricity, Industry and Class in South Africa,* St. Antony's Series (London: Macmillan, 1984).

12. Ben Fine and Zavareh Rustomjee, *The Political Economy of South Africa: From Minerals-Energy Complex to Industrialisation* (London: C. Hurst, 1996).

13. Nancy L. Clark, *Manufacturing Apartheid: State Corporations in South Africa* (New Haven, CT: Yale University Press, 1994); This is also the core insight of Bill Freund's latest book, *Twentieth-Century South Africa: A Developmental History* (Cambridge: Cambridge University Press, 2018).

14. Clark, *Manufacturing Apartheid,* 43.

15. Timothy Mitchell, *Carbon Democracy: Political Power in the Age of Oil* (London: Verso, 2011).

16. Cooper, *Africa since 1940,* 85. Cooper sees the period of development as spanning 1940–73.

17. J. Tischler, *Light and Power for a Multiracial Nation: The Kariba Dam Scheme in the Central African Federation* (New York: Macmillan, 2013).

18. Allen F. Isaacman and Barbara S. Isaacman, *Dams, Displacement and the Delusion of Development: Cahora Bassa and Its Legacies in Mozambique, 1965—2007* (Athens: Ohio University Press, 2013), 4.

19. Crawford Young, *The Rise and Decline of the Zairian State* (Madison: University of Wisconsin Press, 1985).

20. Jeffrey Herbst, *States and Power in Africa: Comparative Lessons in Authority and Control* (Princeton, NJ: Princeton University Press, 2000).

21. Harry Verhoeven, *Water, Civilization and Power in Sudan: The Political Economy of Military-Islamist State-Building* (Cambridge: Cambridge University Press, 2015); it is important to note that studies on the question of state-society linkages have emphasized the importance of linkages that transcend physical proximity and distance: Catherine Boone, *Political Topographies of the African State: Territorial Authority and Institutional Choice* (Cambridge: Cambridge University Press, 2003).

22. Iginio Gagliardone, *The Politics of Technology in Africa: Communication, Development, and Nation-Building in Ethiopia* (Cambridge: Cambridge University Press, 2016).

23. Saul Dubow, for instance, has highlighted the role scientific institutions played in unifying the four warring provinces in South Africa to create the Union of South Africa in 1910. Saul Dubow, *A Commonwealth of Knowledge: Science, Sensibility, and White South Africa, 1820–2000* (Oxford: Oxford University Press, 2006).

24. Keith Breckenridge, "Verwoerd's Bureau of Proof: Total Information in the Making of Apartheid," *History Workshop Journal* 59, no. 1 (20 March 2005): 83–108.

25. William Beinart and Saul Dubow, *The Scientific Imagination in South Africa: 1700 to the Present* (Cambridge: Cambridge University Press, 2021).

26. Thomas Parke Hughes, "The Seamless Web: Technology, Science, Etcetera, Etcetera," *Social Studies of Science* 16, no. 2 (1986): 281–92.

27. Gabrielle Hecht, *The Radiance of France: Nuclear Power and National Identity after World War II* (Cambridge, MA: MIT Press, 1998), 15.

28. Timothy Mitchell, *Rule of Experts: Egypt, Techno-Politics, Modernity* (Berkeley: University of California Press, 2002).

29. Antina von Schnitzler, *Democracy's Infrastructure: Techno-Politics and Protest after Apartheid* (Princeton, NJ: Princeton University Press, 2016), 11.

30. Michel Callon and Bruno Latour, "Unscrewing the Big Leviathan: How Actors Macro-Structure Reality and How Sociologists Help Them to Do So," in *Advances in Social Theory and Methodology: Toward an Integration of Micro-and Macro-Sociologies,* ed. Karin Knorr Cetina and Aaron Victor Cicourel (Boston: Routledge and Kegan Paul, 1981), 277–303.

31. Bruno Latour, *The Pasteurization of France* (Cambridge, MA: Harvard University Press, 1993), 65.

32. Susan Leigh Star and James R. Griesemer, "Institutional Ecology, 'Translations' and Boundary Objects: Amateurs and Professionals in Berkeley's Museum of Vertebrate Zoology, 1907–39," *Social Studies of Science* 19, no. 3 (1989): 387–420.

33. Brian Larkin, "The Politics and Poetics of Infrastructure," 329.

34. Jon Inggs and Stuart Jones, "An Overview of the South African Economy in the 1980s," *South African Journal of Economic History* 9, no. 1 (1 September 1994): 1–18.

35. Ben Fine, "Privatisation and the RDP: A Critical Assessment," *Transformation* 27 (1995): 2–23.

36. Willem Johannes De Villiers, *Report of the Commission of Inquiry Into the Supply of Electricity in the Republic of South Africa* (Pretoria: Government Printer, 1984).

37. Cooper, *Africa since 1940.*

38. Nicolas van de Walle, *African Economies and the Politics of Permanent Crisis, 1979–1999* (Cambridge: Cambridge University Press, 2001).

39. Von Schnitzler, *Democracy's Infrastructure.*

40. Deborah Posel, "Language, Legitimation and Control: The South African State after 1978," *Social Dynamics* 10, no. 1 (1 June 1984): 1.

41. Rajesh Venugopal, "Neoliberalism as Concept," *Economy and Society* 44, no. 2 (3 April 2015): 182.

42. Von Schnitzler, *Democracy's Infrastructure.*

43. Pierre Bourdieu, "Utopia of Endless Exploitation: The Essence of Neoliberalism," trans. Jeremy J. Shapiro, *Le Monde diplomatique* (English ed.), December 1998.

44. David Harvey, *A Brief History of Neoliberalism* (Oxford: Oxford University Press, 2007), 7.

45. James Ferguson, "The Uses of Neoliberalism," *Antipode* 41 (1 January 2010): 166–84.

46. Stephen J. Collier, *Post-Soviet Social: Neoliberalism, Social Modernity, Biopolitics* (Princeton, NJ: Princeton University Press, 2011).

47. Thomas Parke Hughes, *Networks of Power: Electrification in Western Society, 1880–1930,* Reprint edition (Baltimore, MD: Johns Hopkins University Press, 1993).

48. Stephen Graham and Simon Marvin, *Splintering Urbanism: Networked Infrastructures, Technological Mobilities and the Urban Condition* (London: Routledge, 2002).

49. Von Schnitzler, *Democracy's Infrastructure;* Collier, *Post-Soviet Social.*

50. Cooper, *Africa since 1940.*

51. Robert Bates, *When Things Fell Apart: State Failure in Late-Century Africa* (Cambridge: Cambridge University Press, 2008).

52. Daniel Mains has made a similar point in relation to ubiquitous, ongoing construction projects around the world: *Under Construction: Technologies of Development in Urban Ethiopia* (Durham, NC: Duke University Press, 2019), 12.

53. Ministry of Finance, Government of India, *The BRICS Report: A Study of Brazil, Russia, India, China, and South Africa with Special Focus on Synergies and Complementarities* (New Delhi: Oxford University Press, 2012).

54. Michel Callon, Pierre Lascoumes, and Yannick Barthe, *Acting in an Uncertain World: An Essay on Technical Democracy,* trans. Graham Burchell (Cambridge, MA: MIT Press, 2009); Andrew Barry, *Material Politics: Disputes Along the Pipeline* (Chichester, UK: John Wiley & Sons, 2013).

55. Sheila Jasanoff, *Designs on Nature: Science and Democracy in Europe and the United States* (Princeton, NJ: Princeton University Press, 2007), 19.

CHAPTER 1: THE UNLIKELY EXPLOITATION OF THE WATERBERG

1. Hansard (Record of Parliamentary Proceedings, published by Parliament of South Africa), 23 March 1961, col 3591.

2. Charles H. Feinstein, *An Economic History of South Africa: Conquest, Discrimination, and Development* (Cambridge: Cambridge University Press, 2005), 221.

3. Hansard, 18 May 1962, col 5939.

4. Hansard, 15 January 1966, col 74–75.

5. Hansard, col 75.

6. Daryl Glaser, "The State, Capital and Industrial Decentralisation Policy in South Africa, 1932–1985" (master's thesis, University of the Witwatersrand, 1988).

7. Thomas Parke Hughes, "Technological Momentum," in *Does Technology Drive History? The Dilemma of Technological Determinism,* ed. Michael L Smith and Leo Marx (Cambridge, MA: MIT Press, 1994), 101–13.

8. Shadreck Chirikure, *Metals in Past Societies: A Global Perspective on Indigenous African Metallurgy* (Cham, Switzerland: Springer 2015), 64. Chirikure demonstrates that tree felling led to ecological destruction, such as in Mali.

9. Nancy L. Clark, *Manufacturing Apartheid: State Corporations in South Africa* (New Haven, CT: Yale University Press, 1994).

10. Alice Jacobs, *South African Heritage: A Biography of H. J. Van Der Bijl* (Pietermaritzburg, South Africa: Shuter & Shooter, 1948).

11. Jacobs.

12. Bill Freund, *Twentieth-Century South Africa: A Developmental History* (Cambridge: Cambridge University Press, 2018), 37.

13. G. R. Bozzoli, *Forging Ahead: South Africa's Pioneering Engineers* (Johannesburg: Witwatersrand University Press, 1997), 78.

14. Jacobs, *South African Heritage*, 122.

15. Foreman Bandama, Shadreck Chirikure, and Simon Hall, "Ores Sources, Smelters and Archaeometallurgy: Exploring Iron Age Metal Production in the Southern Waterberg, South Africa," *Journal of African Archaeology* 11, no. 2 (2013): 243–67.

16. F. Meyer, "Presidential address to the Associated Scientific and Technical Societies of South Africa," 15 November 1961, National Archives of South Africa (hereafter referred to as NASA), MES 243 H4/12/2.

17. Richard Mendelsohn, *Sammy Marks: The Uncrowned King of the Transvaal* (Cape Town: David Philip, 1991).

18. Jan Smuts, "South Africa's Social and Economic Planning Council: General Smuts's Plans for a Better and Greater Southern Africa," *Journal of the Royal African Society* 41, no. 165 (October 1942): 231–33.

19. Social and Economic Planning Council (South Africa), *Public Works Programme and Policy,* ed. Hendrik Johannes Van Eck, South Africa. Report, No. 10 (Cape Town: Cape Times [for Government Printer], 1946), 2.

20. Social and Economic Planning Council (South Africa), 10.

21. Albert O. Hirschman, *Development Projects Observed* (Washington, DC: Brookings Institution, 1967).

22. Saul Dubow, *A Commonwealth of Knowledge: Science, Sensibility, and White South Africa, 1820–2000* (Oxford: Oxford University Press, 2006).

23. Dubow, 258

24. Dubow, 251.

25. Hansard, 6 February 1974, col 219.

26. Hansard, 16 May 1969, col 6133.

27. J. B. Mills, South African ambassador in Pretoria, to the Secretary for Foreign Affairs, 11 January 1972, NASA, MES 239 H4/12/2.

28. Cape Midlands Development Association, *Memorandum: Growth Points and the Siting of the Fourth Iscor,* 15 March 1972, NASA, MES 239 H4/12/2.

29. Cape Midlands Development Association.

30. "Japanese Seek Ore Berth Policy Switch," extract from *Eastern Province Herald,* 10 March 1972, NASA, MES 239 H4/12/2.

31. "Japanese Slate Apartheid," extract from *Eastern Province Herald,* 2 February 1972, NASA, MES 239 H4/12/2.

32. "Iscor Survey," *Business Times,* 26 November 1972, NASA, MES 239 H4/12/2.

33. Ben Alberts, "Presidential Address: Planning the Utilisation of South Africa's Coal Reserves," *Journal of South African Institute of Mining and Metallurgy* 87, no. 11 (November 1987): 388.

34. Ruth Edgecombe and Bill Guest, "The Natal Coal Industry in the South African Economy, 1910–1985," *South African Journal of Economic History* 2, no. 2 (1987): 52.

35. Bill Guest, "Commercial Coal-Mining in Natal: A Centennial Appraisal," *Natalia* 18 (1998): 55.

36. Stephen Sparks, "Apartheid Modern: South Africa's Oil from Coal Project and the History of a Company Town" (PhD diss., University of Michigan, 2012); Gabrielle Hecht, *Being Nuclear: Africans and the Global Uranium Trade* (Cambridge, MA: MIT Press, 2012); Stephen Gelb, ed., *South Africa's Economic Crisis* (Cape Town: David Philip, 1991); Andrew Marquard, "The Origins and Development of South African Energy Policy" (PhD diss., University of Cape Town, 2006), 77, http://webdav.uct.ac.za/depts/erc /Research/publications/06Marquard%20PhD%20Thesis.pdf.

37. J. C. Heunis to P. J. G. Koornhof, "Onluste by Hlobane Northfield Steenkoolmyn," 20 March 1973, NASA, MES 233 H4/12/3.

38. Heunis to Koornhof.

39. D. W. Horsfall, "A General Review of Coal Preparation in South Africa," in *Journal of the South African Institute of Mining and Metallurgy* 80, no. 8 (August 1980): 259.

40. Michael Deats, interview with author, 2 September 2013, Woodmead, Gauteng.

41. Deats, interview with author.

42. Deats, interview with author.

43. J. P. Coetzee to G. Steyn, "Steenkool Posisie in Suid-Afrika—Yskor se Siening," 13 May 1974, NASA, MES 234 H4/12.

44. Yskor to J. C. Heunis, "Yskor en die Private Nywerheid in die Republiek van Suid Afrika—Quo Vadis?," 17 March 1975, NASA, MES 245 H4/12/3.

45. Minutes of Board Meeting of the South African Iron and Steel Industrial Corporation Limited, 14 July 1976, NASA, MPP 44 A3/10/9.

46. T. F. Muller to S. L. Muller, 16 May 1974, NASA, MPP 47 A3/10/9.

47. Minutes of Board Meeting of the South African Iron and Steel Industrial Corporation Limited, 23 July 1975, NASA, MPP 44 A3/10/9.

48. Minutes of Board Meeting of the South African Iron and Steel Industrial Corporation, 25 May 1977, NASA, MPP 49 A3/10/9.

49. J. P. Coetzee to J. C. Heunis, 23 March 1977, NASA, MPP 24 A3/10/2.

50. Deats, interview with author.

51. Eugene Marais, *The Road to Waterberg and Other Essays* (Cape Town: Human & Rousseau, 1972), 10.

52. Isabel Hofmeyr, "Turning Region into Narrative: English Storytelling in the Waterberg" (paper presented at conference, The Making of Class, University of the Witwatersrand, History Workshop, 1987), 11.

53. Marais, *The Road to Waterberg and Other Essays.*

54. Marais, 10.

55. P. E. Hall, "A Preliminary Note on the Waterberg Coalfield as a Possible Source of Coking Coal," *Journal of the Chemical, Metallurgical and Mining Society of South Africa* (October 1945), 124–32.

56. Deats, interview with author.

57. "Grootgeluk Funding," n.d., MPP 44 A3/10/9.

58. "Memorandum, Investment in the Steel Sector, the World Bank's Model and South Africa's Co-operation," 9 June 1976, NASA, MPP 23/A3/10/1.

59. Iscor Annual Report, 1978, NASA, MPP 50 A3/10/9.

60. "Memorandum."

61. Minutes of Board Meeting of the South African Iron and Steel Industrial Corporation, 16 July 1976, NASA, MPP 44 A3/10/9.

62. Minutes of Board Meeting of the South African Iron and Steel Industrial Corporation, 25 February 1976, NASA, MPP 44 A3/10/9.

63. Minutes of Board Meeting of the South African Iron and Steel Industrial Corporation, 26 April 1978, NASA, MPP 50 A3/10/9.

64. Minutes of Board Meeting of the South African Iron and Steel Industrial Corporation, 25 May 1977, NASA, MPP 49 A3/10/9.

65. Minutes of Board Meeting of the South African Iron and Steel Industrial Corporation, 28 September 1977, NASA, MPP 50/A3/10/9.

66. Minutes of Board Meeting of the South African Iron and Steel Industrial Corporation, 27 July 1977, NASA, MPP 50 A3/10/9.

67. Deats, interview with author.

68. Deats, interview with author.

69. Minutes of Board Meeting of the South African Iron and Steel Industrial Corporation, 28 September 1977, NASA, MPP 50 A3/10/9.

70. Joe Meyer, interview with author, 17 March 2015, Onverwacht.

71. Minutes of Board Meeting of the South African Iron and Steel Industrial Corporation, 30 August 1978, NASA, MPP 50 A3/10/9.

72. Meyer, interview with author.

CHAPTER 2: THE TAMING OF THE WATERBERG

1. Shula Marks and Stanley Trapido, "Lord Milner and the South African State," *History Workshop,* no. 8 (1979): 50–80.

2. Dan O'Meara, *Forty Lost Years: The Apartheid State and the Politics of the National Party, 1948–1994* (Johannesburg: Ravan, 1996), 377.

3. Deborah Posel, "Language, Legitimation and Control: The South African State after 1978," *Social Dynamics* 10, no. 1 (1 June 1984): 1–16.

4. "PW and the Gamble That Failed," *Sunday Express,* 26 November 1978.

Notes to Pages 38–44 ⁓ 137

5. "Ellisras Ontwikkeling," 10 September 1965, NASA, RNH 105 NH5/55.
6. Ponk Ellis, interview with author, August 2013, Lephalale.
7. Faeeza Ballim, "The Un-Making of the Group Areas Act: Local Resistance and Commercial Power in the Small Town of Mokopane," *South African Historical Journal* 69, no. 4 (2 October 2017): 568–82.
8. Ellis, interview with author.
9. Sam Sekati, interview with author, 4 August 2014, Lephalale.
10. Johannes Mfisa, group interview with Hellen Kekae, Hendrik Ndebele, April Selema, and Kgantshi Makubela (translator), Marapong, Lephalale, 25 August 2015.
11. Sekati, interview with author.
12. Willie Loots, interview with author, 6 August 2014, Lephalale.
13. The term "Northern Sotho" refers to a language spoken by people in communities of the northern parts of South Africa.
14. Bantoesakekommisaris to Sekr van B. A. & O., 9 May 1968; NASA, BAO 3182 C39/1839.
15. Loots, interview with author.
16. Faeeza Ballim, "The Pre-History of South African 'Neo-Liberalism': The Rise and Fall of Co-Operative Farming on the Highveld," *Journal of Southern African Studies* 41, no. 6 (2 November 2015): 1239–54, https://doi.org/10.1080/03057070.2015.1093835.
17. William Beinart, *Twentieth-Century South Africa* (Oxford: Oxford University Press, 2001).
18. Ballim, "The Pre-History of South African 'Neo-Liberalism.'"
19. "Potensial van Ellisras," November 1964, NASA, ACE 50 TS/18/18/1.
20. "Potensial van Ellisras," 4.
21. Johan Pistorius, interview with author, March 2015, Lephalale.
22. Pistorius, interview with author.
23. Nqobile Zulu, "An Analysis of the Post 1980s Transition from Pastoral to Game Farming in South Africa: A Case Study of the Marico District" (PhD diss., University of the Witwatersrand, 2015).
24. Timothy Keegan, *Rural Transformations in Industrializing South Africa: The Southern Highveld to 1914* (Johannesburg: Ravan, 1986), 103.
25. Stadsgebiedekommisaris to die Ondersekretaris, Native Affairs Department, *Verslag: Dorp Ellisras: Distrik Waterberg,* 19 June 1959, NASA, NTS 4572 1219/313.
26. H. J. Combrink to the Head regional health official, 30 October 1957, NASA, CDB 3209 PB4/2/2/1974.
27. This official title includes a word that is today considered derogatory. This word has been placed in quotation marks to highlight this fact.
28. "Proclamation," *Province of Transvaal Official Gazette,* 7 December 1960, NASA, CDB 3209 PB 4/2/2/1794.
29. L. J. Botha to Direkteur van Plaaslike Bestuur [Director of Local Management], November 1965, NASA, CDB 3209 PB4/2/2/1974.

30. "Summary of Problems in Ellisras Town in Connection with Water, Cemeteries, Land Fill Ground and Location Ground," 4 April 1966, NASA, TRB 2/4/57 G31/13/0.

31. M. J. Deats to the Secretary, Tvl Board for the Development of Peri-urban areas, 29 October 1976.

32. I. D. Potgieter to the Konsult Mynbou-Ingenieur, Yskor, "Water Provision to the Suggested Town Ellisras," 16 August 1976, TRB 2/1/177 61/1/480.

33. Head of Department of Local Areas to Louis Botha and Potgieter, 25 April 1977, TRB 2/1/174 61/1/480.

34. Head of Department of Local Areas to Botha; Louis Botha and Potgieter to the Head of the Department of Local Areas, 7 April 1977.

35. Beinart, *Twentieth-Century South Africa.*

36. Mosa Phadi and Joel Pearson, *We Are Building a City: The Struggle for Self-Sufficiency in Lephalale Local Municipality* (Johannesburg: Public Affairs Research Institute, 2018), https://pari.org.za/7221-2/.

37. Joe Meyer, interview with author, 17 March 2015, Onverwacht.

38. Dries de Ridder, interview with author, April 2015, Lephalale.

39. Meyer, interview with author.

40. "Ellisras Vergadering," 3 July 1975, NASA, TRB 2/4/57 G31/13/0.

41. "Official Development of the Town," Special Meeting, Ellisras, 3 July 1975, TRB 2/4/57 G31/13/0.

42. Dorperaad die Voorsitter, "Ellisras dorpsbeplanningskema," 7 April 1987, NASA, CDB 9987 PB4/9/2.

43. This official title includes a word that is today considered derogatory. This word has been placed in quotation marks at first mention to highlight this fact.

44. "Stationing a Permanent Health Inspector: Ellisras Plaaslike Gebiedskomitee," 1 April 1966, NASA, TRB 2/4/57 G31/13/0.

45. Hoofmediese Gesondheidsbeampte to Hoofbantoesakekommissarie, 20 March 1976, NASA, TRB 2/4/57 G31/13/0.

46. Hoofmediese Gesondheidsbeampte to Hoofbantoesakekommissarie.

47. Head Bantu Commissioner for Native Affairs to the Head Medical Health official of the Transvaal, Peri-Urban Areas Board, 22 September 1976, NASA, TRB 2/4/57 G31/13/0.

48. "Memorandum: Aanonafgehandelde sake," 4 May 1976, NASA, TRB 2/4/57 G31/13/0.

49. "Gesamentlike aksieprogram teen inflasie: Invordering van bantoebelasting," 7 October 1976, NASA, TRB 2/4/57 G31/13/0.

50. "Blacks Hold the Key to 'Constellation', PM Told," *Rand Daily Mail,* 30 August 1979.

51. Saul Dubow, *Apartheid, 1948–1994* (Oxford: Oxford University Press, 2014), 117.

52. "Tvl Border Areas Now a 'White Border Post,'" *Pretoria News,* 4 February 1982.

53. "Massive Exodus from Border Farmlands," *The Citizen* (Johannesburg), 21 August 1978.

54. "Grensplase Ontvolk: Depopulation of Border Farms," *Die Transvaler,* 26 May 1978.

55. "Police Kill a Suspected Terrorist," *The Star* (Gauteng), 7 August 1984.

56. "Idle Farms Get Blame for Mines," *The Star* (Gauteng), 7 January 1986.

57. "Government Approves Plan to Boost Border Security," *The Citizen* (Johannesburg), 12 December 1984; "R34 Budgeted to Stabilise Transvaal Border Areas," *The Star* (Gauteng), 13 December 1984.

58. "SA's Frontline Dwindles," *Sunday Tribune,* 20 January 1980.

59. M. N. Ramodike, "Address by the New Chief Minister," Lebowa legislative assembly, Lebowakgomo, 21 October 1987, NASA, MSB 706 3/5/10/3/8.

60. "20,000 Are Banished to Dumping Grounds," *Sunday Express,* 29 July 1979.

61. James C. Scott, *Seeing Like a State: How Certain Schemes to Improve the Human Condition Have Failed* (New Haven, CT: Yale University Press, 1998).

CHAPTER 3: ESKOM AND THE TURNING OF THE TIDE

1. See Franziska Rueedi, *The Vaal Uprising of 1984 and the Struggle for Freedom in South Africa* (Woodbridge, Suffolk: James Currey, 2021).

2. Robert Davies and Dan O'Meara, "Total Strategy in Southern Africa: An Analysis of South African Regional Policy Since 1978," *Journal of Southern African Studies* 11, no. 2 (1 April 1985): 183–211,

3. Ian McRae, *The Test of Leadership: 50 Years in the Electricity Supply Industry of South Africa* (Muldersdrift, South Africa: EE Publishers, 2006), 31.

4. McRae, 31. Ian McRae served as the first manager of the Central Generation Undertaking (CGU).

5. Steve R. Conradie and L. J. M. Messerschmidt, *A Symphony of Power: The Eskom Story* (Johannesburg: Chris Van Rensburg, 2000), 142.

6. Conradie and Messerschmidt.

7. Anton Eberhard, "The Political Economy of Power Sector Reform in South Africa," in *The Political Economy of Power Sector Reform: The Experiences of Five Major Developing Countries,* ed. David G. Victor and Thomas C. Heller (Cambridge: Cambridge University Press, 2007).

8. McRae, *The Test of Leadership,* 33.

9. Hansard, 3 June 1968, col 6411; He cited the government-commissioned Borckenhagen Report, released in 1968: *White Paper on the Reports of the Committee of Enquiry into the Financial Relations Between the Central Government, the Provinces and the Local Authorities: The Borckenhagen Reports* (Pretoria: Republic of South Africa, 1971).

10. McRae, *The Test of Leadership,* 33.

11. Conradie and Messerschmidt, *A Symphony of Power.*

12. Conradie and Messerschmidt.

13. "Memorandum: Updated Programme for Expanding Eskom's Generating Capacity," 24 April 1978, Eskom Corporate Archives, 2L-10 KS 258.

14. So significant has the relationship between the two been that political economist Ben Fine has described the South African economy as characterized by the dominance of the "minerals-energy complex," which involves activities in the minerals and energy sectors. Ben Fine, "The Minerals-Energy Complex Is Dead: Long Live the MEC" (paper presented to Amandla Colloquium, Cape Town, 15 October 2008), https://eprints.soas.ac.uk/5617/1/MineralEnergyComplex.pdf.

15. Minutes of meeting of Electricity Supply Commission, 5 March 1980, Eskom Heritage Department Archives, Minute Book no. 18.

16. Minutes of meeting of Electricity Supply Commission, 5 March 1980.

17. "Eskom's Tap-Dance," *Financial Mail,* 18 August 1989.

18. Minutes of meeting of Electricity Supply Commission, 15 February 1977, Eskom Corporate Archives, 2L-10 KS 257.

19. Minutes of meeting of Electricity Supply Commission, 15 February 1977; minutes of meeting of Electricity Supply Commission, 27 September 1977, Eskom Corporate Archives, 2L-10 KS 258.

20. Minutes of meeting of Electricity Supply Commission, 25 October 1977, Eskom Corporate Archives, 2L-10 KS 258.

21. Minutes of meeting of Electricity Supply Commission, 15 October 1977.

22. Minutes of meeting of Electricity Supply Commission, 24 January 1978, Eskom Corporate Archives, 2L-10 KS 258.

23. Minutes of special tender board meeting, 23 August 1979, Eskom Corporate Archives, O9 KS 242.

24. Minutes of meeting of Electricity Supply Commission, 13 June 1978, Eskom Corporate Archives, 2L-10, KS 259.

25. Minutes of meeting of Electricity Supply Commission, 22 April 1980, Eskom Corporate Archives, Microfilms.

26. Minutes of meeting of Electricity Supply Commission, 12 October 1976, Eskom Corporate Archives, 2L-10 KS 257.

27. "Introducing CIGRE," CIGRE, accessed 4 November 2022, https://www.cigre.org/GB/about/introducing-cigre.

28. Minutes of meeting of Electricity Supply Commission, 12 October 1976.

29. Minutes of special tender board meeting, 23 August 1979.

30. Minutes of meeting of Electricity Supply Commission, 5 May 1982, Eskom Heritage Archives, Minute Book 19.

31. "Bonn Guarantees R341-m for Second SA Power Scheme," *The Star* (Gauteng), 8 October 1981.

32. Paul N. Edwards and Gabrielle Hecht, "History and the Technopolitics of Identity: The Case of Apartheid South Africa," *Journal of Southern African Studies* 36, no. 3 (1 January 2010): 637. It should also be kept in mind that German engineering companies have historically been important to the state corporations, and many of the prominent engineers were trained at German universities. G. R. Bozzoli, *Forging Ahead: South Africa's Pioneering Engineers* (Johannesburg: Witwatersrand University Press, 1997), 172.

33. "Matimba: Project of the Year '87," special issue, *Engineering Week,* November 1987.
34. "Matimba," 1.
35. "Matimba," 1.
36. "A Giant Step for Eskom," in "Matimba," 6.
37. Minutes of meeting of Electricity Supply Commission, 1 April 1980, Eskom Heritage archives.
38. J. L. Rothman to General Manager, "Memorandum: Dry Cooling of Turbines with Reference to Future Eskom Applications," 17 January 1980, Eskom Corporate Archives, Microfilms.
39. Minutes of meeting of Electricity Supply Commission, 22 April 1980, Eskom Heritage Archives.
40. Minutes of meeting of Electricity Supply Commission, 1 April 1984, Eskom Heritage Archives, Minute Book 19.
41. Interview with author (anonymous), 25 June 2015, Johannesburg Country Club.
42. Interview with author (anonymous), 25 June 2015.
43. Rothman, "Memorandum."
44. Minutes of meeting of Electricity Supply Commission, 27 August 1981, Eskom Corporate archives, Microfilms.
45. Minutes of meeting of Electricity Supply Commission, 7 October 1981, Eskom Corporate archives, Microfilms.
46. Minutes of meeting of Electricity Supply Commission, 5 August 1981, Eskom Corporate Archives, Microfilms.
47. "In the Pound Seats," *Financial Mail,* 14 May 1989.
48. "Local Boilers for Matimba," 19.
49. Minutes of meeting of Electricity Supply Commission, 14 July 1981, Eskom Corporate Archives, Microfilms; Minutes of meeting of Electricity Supply Commission, 7 July 1981, Eskom Corporate Archives, Microfilms.
50. Frederick Cooper, *Africa since 1940: The Past of the Present,* New Approaches to African History (Cambridge: Cambridge University Press, 2002), 34.
51. "The Recommendations of the Riekert Commission Are Tabled," South African History Online, accessed 12 November 2016, http://www.sahistory.org.za/dated-event/recommendations-riekert-commission-investigate-employment-conditions-black-workers-are-t.
52. Personnel Manager to the Senior General Manager, "Eskom Housing for Non-White Employees," 24 September 1980, Eskom Corporate Archives, Microfilms.
53. Personnel Manager to the Senior General Manager.
54. J. G. G. Strydom to die Sekrataris, "Ellisras Distrikontwikkelingsvereniging," 20 November 1986, NASA, RLA 745 20/5/E36/1.
55. J. G. G. Strydon to die Sekrataris.
56. J. P. Deetleefs to Mynbestuurder, "Grootgeluk Steenkoolmyn," 4 September 1986, NASA, RLA 745 20/5/E36/1.

57. "Ellisras Kommentaar," *Die Kwevoel,* 18 September 1987, NASA, CDB 2138 PB3/13/2/152.
58. W. A. Lewis, Secretary of the Distrik Landbou, "Memorandum insake swartdorp te Ellisras," n.d., NASA, RLA 745 20/5/E36/1.
59. A. B. Treurnicht to Minister J. C. Heunis, 4 December 1987, NASA, RLA 745 20/5/E36/1.
60. Mogol Kommando, "Memorandum: Voorgestelde swartdorp in die Marapong area," n.d., NASA, RLA 745 20/5/E36/1.
61. F. W. de Klerk to Minister J. C. Heunis, 15 July 1987, NASA, RLA 745 20/5/E36/1.

CHAPTER 4: CONTESTED NEOLIBERALISM

1. Jon Inggs and Stuart Jones, "An Overview of the South African Economy in the 1980s," *South African Journal of Economic History* 9, no. 1 (1 September 1994): 1–18.
2. Frederick Cooper, *Africa since 1940: The Past of the Present,* New Approaches to African History (Cambridge: Cambridge University Press, 2002).
3. Nicolas van de Walle, *African Economies and the Politics of Permanent Crisis, 1979–1999* (Cambridge: Cambridge University Press, 2001), 84.
4. Van de Walle.
5. Michel Foucault, *The Birth of Biopolitics: Lectures at the Collège de France, 1978–1979* (New York: Picador, 2010).
6. Dan O'Meara, *Forty Lost Years: The Apartheid State and the Politics of the National Party, 1948–1994* (Johannesburg: Ravan, 1996), 377.
7. Deborah Posel, "Language, Legitimation and Control: The South African State after 1978," *Social Dynamics* 10, no. 1 (1 June 1984): 1–16.
8. Antina von Schnitzler, *Democracy's Infrastructure: Techno-Politics and Protest after Apartheid* (Princeton, NJ: Princeton University Press, 2016), 43.
9. Von Schnitzler, 54.
10. Faeeza Ballim, "The Pre-History of South African 'Neo-Liberalism': The Rise and Fall of Co-Operative Farming on the Highveld," *Journal of Southern African Studies* 41, no. 6 (2 November 2015): 1239–54.
11. Vishnu Padayachee, "Private International Banks, the Debt Crisis and the Apartheid State, 1982–1985," *African Affairs* 87, no. 348 (1 July 1988): 363.
12. Padayachee, 367.
13. "Eskom Annual Report," 1986, 7. https://www.eskom.co.za/sites/heritage/Pages/Annual-Reports.aspx.
14. Stephen J. Collier, *Post-Soviet Social: Neoliberalism, Social Modernity, Biopolitics* (Princeton, NJ: Princeton University Press, 2011), 243.
15. Anton Eberhard, "From State to Market and Back Again: South Africa's Power Sector Reforms," *Economic and Political Weekly* 40, no. 50 (10 December 2005): 5309–17.
16. O'Meara, *Forty Lost Years,* 399. O'Meara writes that Thatcher wrote to F. W. de Klerk, warning that "she would not be able to sustain her anti-sanctions stance forever."

Notes to Pages 71–76 ～ 143

17. *White Paper on Privatization and Deregulation in the Republic of South Africa* (Pretoria: Government Printer, South Africa, 1987).
18. *White Paper on Privatization and Deregulation in the Republic of South Africa,* 4.
19. Hansard, 10 May 1990, col 8623.
20. "De Villiers Wants to Make Capital Work," *Business Day,* 14 March 1989.
21. "Transfer Pending," *Leadership SA,* 1989.
22. "The Man from Gencor," *Rand Daily Mail,* 17 June 1982.
23. "De Villiers Wants to Make Capital Work."
24. "De Villiers Wants to Make Capital Work."
25. Willem Johannes De Villiers, *Report of the Commission of Inquiry into the Supply of Electricity in the Republic of South Africa* (Pretoria: Government Printer, 1984).
26. "De Villiers Wants to Make Capital Work."
27. Willem Johannes De Villiers, "Privatization of Eskom, SA Vervoerdienste, Pos En Telekommunikasie," 25 May 1988, NASA, MPP 2 1/3/1.
28. "Debunking the Eskom Myth," *Financial Times,* 2 November 1990.
29. "De Villiers Wants to Make Capital Work."
30. Hansard, assembly debates, 10 May 1990, 8626.
31. Hansard, 8685.
32. Allen F. Isaacman and Barbara S. Isaacman, "Extending South Africa's Tentacles of Empire: The Deterritorialisation of Cahora Bassa Dam," *Journal of Southern African Studies* 41, no. 3 (June 2015): 541–60.
33. Anton Eberhard, "The Political Economy of Power Sector Reform in South Africa" (working paper WP-06, Program on Energy and Sustainable Development, Stanford University, April 2004), 8.
34. "SA Unions Set for Big Battle with Privatization," *The Star* (Gauteng), 29 March 1990.
35. Hansard, 6 May 1992, col 6328.
36. Grove Steyn, "Investment and Uncertainty: Historical Experience with Power Sector Investment in South Africa and Its Implications for Current Challenges" (working paper prepared for the Management Programme in Infrastructure Reform and Regulation at the Graduate School of Business, University of Cape Town, 15 March 2006), 41.
37. "Transfer Pending," *Leadership SA,* 1989.
38. P. Bond, *Elite Transition: From Apartheid to Neoliberalism in South Africa* (London: Pluto, 2000).
39. Eberhard, "From State to Market and Back Again," 5314.
40. *White Paper on the Energy Policy of the Republic of South Africa* (Pretoria: Department of Minerals and Energy, 1998).
41. *White Paper on the Energy Policy of the Republic of South Africa,* 7.
42. William Mervin Gumede, *Thabo Mbeki and the Battle for the Soul of the ANC* (London: Zed, 2007), 119.
43. "Mbeki Apologises for SA Power Cuts," *Independent Online,* 12 December 2007.

44. Clive Le Roux, interview with author, 18 February 2015, Megawatt Park, Gauteng.
45. Le Roux, interview with author.
46. Le Roux, interview with author.
47. Eskom Conversion Act [No. 13 of 2001], Government Gazette, 3 August 2001.
48. "Eskom Conversion Bill: Cosatu Input; Department Briefing on Additional Amendments," Parliamentary Monitoring Group, 12 June 2001, https://pmg.org.za/committee-meeting/596/.
49. Eberhard, "From State to Market and Back Again."
50. Le Roux, interview with author.
51. Le Roux, interview with author.
52. Le Roux, interview with author. The problem of the cracks in the boiler exhaust ducts had arisen years earlier: P. de W. la Grange to Matimba Power station manager, "Flue gas exhaust ducts," 6 June 1991, Matimba Power Station Document Centre, 4066123.
53. Interview with author (anonymous), 25 June 2015, Johannesburg Country Club.
54. Le Roux, interview with author.
55. "1997 National Productivity Awards," Matimba Power Station Document Centre, Lephalale, 4075785.
56. "1997 National Productivity Awards."
57. "1997 National Productivity Awards."
58. "1997 National Productivity Awards."
59. Le Roux, interview with author; the identity of this individual will be protected and will be referred to here as Andile Williams.
60. Le Roux, interview with author.
61. Bheki Nxumalo, interview with author, March 2016, Matimba Power Station, Lephalale.
62. Antina Von Schnitzler, "Travelling Technologies: Infrastructure, Ethical Regimes, and the Materiality of Politics in South Africa," *Cultural Anthropology* 28, no. 4 (1 November 2013): 670–93.
63. Ian McRae, *The Test of Leadership: 50 Years in the Electricity Supply Industry of South Africa* (Muldersdrift, South Africa: EE Publishers, 2006), 13.
64. Le Roux, interview with author.
65. Le Roux, interview with author.
66. Stephen Kekana, interview with author, 16 March 2016, Matimba power station, Lephalale.
67. Kekana, interview with author.

CHAPTER 5: LABOR AND BELONGING IN LEPHALALE

1. Charles H. Feinstein, *An Economic History of South Africa: Conquest, Discrimination, and Development* (Cambridge: Cambridge University Press, 2005).
2. Deborah Posel, *The Making of Apartheid, 1948–1961: Conflict and Compromise* (Oxford: Oxford University Press, 1991).

3. Trade unions and worker protests were also important to decolonization processes elsewhere on the continent as workers agitated for rights to the city and its infrastructural benefits.

4. Andries Bezuidenhout and Sakhela Buhlungu, "From Compounded to Fragmented Labour: Mineworkers and the Demise of Compounds in South Africa," *Antipode* 43, no. 2 (1 March 2011): 237–63. Sakhela Buhlungu and Andries Bezuidenhout argue that the dissolution of the compound system at the mines has fragmented the workforce and posed an organizing challenge, at least for the National Union of Mineworkers.

5. Stephen Sparks, "Apartheid Modern: South Africa's Oil from Coal Project and the History of a Company Town" (PhD diss., University of Michigan, 2012), https://deepblue.lib.umich.edu/handle/2027.42/91528. Dynamics in Lephalale closely resemble those that Stephen Sparks details in his study of the company town of Sasolburg. Sparks describes a situation of labor relations structured on "paternalist" lines rather an on an entirely coercive regime under apartheid.

6. Gabrielle Hecht, *Being Nuclear: Africans and the Global Uranium Trade* (Cambridge, MA: MIT Press, 2012).

7. Jean Leger, "Coal Mining: Past Profits, Current Crisis?," in *South Africa's Economic Crisis,* ed. Stephen Gelb (Cape Town: David Philip, 1991), 143; Sparks, "Apartheid Modern," 215.

8. Sparks, "Apartheid Modern."

9. Saul Dubow, *Apartheid, 1948–1994* (Oxford: Oxford University Press, 2014), 178.

10. Feinstein, *An Economic History of South Africa,* 232.

11. Sparks, "Apartheid Modern," 217. In his study of the "stabilization" of Sasol's company town of Sasolburg, Stephen Sparks demonstrates that, while stabilization coincided with the government's strategy to appease Africans through limited service provision, local-level managers held particular views of the direction of African urban development. Sasol's managers argued that good quality housing for Africans was essential to "developing a middle class black employee group."

12. Sparks.

13. T. Dunbar Moodie, "The Moral Economy of the Black Miners' Strike of 1946," *Journal of Southern African Studies* 13, no. 1 (1986): 3.

14. Donald Reid, "Industrial Paternalism: Discourse and Practice in Nineteenth-Century French Mining and Metallurgy," *Comparative Studies in Society and History* 27, no. 4 (1985): 588.

15. "SA Ready to Stake Billions on Dry-Cooled Answer to Energy Problems," *Rand Daily Mail,* 23 July 1983

16. Stephen Kekana, interview with author, 14 April 2015, Matimba power station.

17. Feinstein, *An Economic History of South Africa,* 242.

18. Hendrik Ndebele, group interview with Johannes Mfisa, Hellen Kekae, Hendrik Ndebele, April Selema, and Kgantshi Makubela (translator), 25 August 2015, Marapong (Lephalale).

19. D. C. U. Conradie, Yskor personeel bestuurder, to Direkteur van Bantoe-Arbeid (Iscor staff manager to director of Bantu Labor), 8 December 1966, NASA, BAO 2541 C31/3/1151/10.

20. Die Hoofbantoesakekommissaris, Pietersburg, to Die Personeelbestuurder, Yskor (head Bantu Affairs commissioner of Pietersburg to the Iscor staff manager), 12 December 1966, NASA, BAO 2541 C31/3/1151/10.

21. Mnr Adendorff to Hoofbantoesakekommisaris (Mr. Adendorff to the head Bantu Affairs commissioner of Pietersburg), 7 June 1971, NASA, BAO 2541 C31/3/1151/10.

22. Inspekteur van Bantoe-Arbeiders, Noordelike Gebied, to Hoofbantoesake-kommisaris (inspector of "bantu" labor, northern territory, to the head Bantu Affairs commissioner of Pietersburg), 20 March 1972, NASA, BAO 2541 C31/3/1151/10.

23. Lazarus Seodisa, interview with author, translated by Kgantshi Makubela, March 2015, Marapong (Lephalale).

24. Seodisa, interview with author.

25. Meyer recalled the cleansing ceremonies that occurred at the excavation site.

26. Seodisa, interview with author.

27. "Medupi Built on Seven Burial Sites," *eNCA* (online), 3 September 2015. https://www.enca.com/south-africa/burial-sites-medupi-was-built.

28. "Medupi Built on Seven Burial Sites."

29. This is discussed in more detail in Chapter Three.

30. Hendrik Ndebele, group interview with Johannes Mfisa, Hellen Kekae, Hendrik Ndebele, April Selema, and Kgantshi Makubela (translator), 25 August 2015, Marapong (Lephalale).

31. Melton Mothoni, interview with author, April 2015, Matimba power station, Lephalale.

32. Stephen Kekana, interview with author, 14 April 2015, Matimba power station.

33. Kekana, interview with author.

34. Kally Forrest, *Metal That Will Not Bend: National Union of Metalworkers of South Africa 1980–1995* (Johannesburg: Witwatersrand University Press, 2011), 17.

35. Dubow, *Apartheid, 1948–1994*, 198.

36. Dubow.

37. Dubow, 18.

38. Kekana, interview with author.

39. Minutes of the Task Group, 19 August 1994, personal collection of M. Mothoni.

40. Minutes of the Task Group, 3.

41. Minutes of Engineering sub BU Forum, 7 September 1994, Matimba power station, personal collection of M. Mothoni.

42. Minutes of a Special Maintenance sub BU Forum, 13 June 1995, personal collection of M. Mothoni.

43. Minutes of a Special Maintenance sub BU forum.

44. Minutes of an Engineering sub BU forum, 7 August 1995, personal collection of M. Mothoni.

45. Minutes of BU forum, 4 May 1995, personal collection of M. Mothoni.

46. Minutes of the Engineering sub BU Forum, 3 August 1994, Matimba power station, personal collection of M. Mothoni.

47. Y. Ballim, R. Omar, and A. Ralphs, "Learning Assessments in South Africa," in *Experiential Learning Around the World: Employability and the Global Economy,* ed. Norman Evans (London: Jessica Kingsley, 2000).

48. Ballim, Omar, and Ralphs.

49. Stephen Kekana, interview with author, 16 March 2016, Matimba power station, Lephalale.

50. Minutes of BU Forum, 29 October 1996, personal collection of M. Mothoni.

51. Minutes of BU Forum, 20 April 1998, personal collection of Stephen Kekana.

52. Interview with author (anonymous), March 2016, Matimba power station, Lephalale.

53. Kekana, interview with author, 16 March 2016.

54. "Bulletin: Industrial Action Issues and Processes," attachment to the minutes of steering committee, 8 August 1997, personal collection of M. Mothoni.

55. Minutes of a BU Forum, 4 September 1997, item 4.4, personal collection of M. Mothoni.

56. Melinde Coetzee, "Recruitment and Selection Procedure," 9 September 1997, personal collection of M. Mothoni.

57. Coetzee.

58. Minutes of a Joint Forum, 4 May 1998, personal collection of Stephen Kekana.

59. Minutes of Joint Forum, 3 August 1998, personal collection of Stephen Kekana.

60. Minutes of a B. U. Forum, 17 January 2000, personal collection of Stephen Kekana.

61. Clive le Roux, interview with author, 18 February 2015, Megawatt Park, Gauteng

62. Le Roux, interview with author.

63. Le Roux, interview with author.

64. Minutes of Joint Forum Meeting, 6 July 1998, personal collection of Stephen Kekana.

65. Minutes of a Joint Forum Meeting, 6 July 1998.

66. Kekana, interview with author, 16 March 2016.

67. Kekana, interview with author, 16 March 2016.

68. Eskom housing policy, 1995.

69. Minutes of a Special Joint Forum, 24 October 1994, personal collection of M. Mothoni.

CHAPTER 6: THE MEDUPI POWER STATION

1. Karan Mahajan, "'State Capture': How the Gupta Brothers Hijacked South Africa Using Bribes Instead of Bullets," *Vanity Fair,* 3 March 2019. https://

www.vanityfair.com/news/2019/03/how-the-gupta-brothers-hijacked
-south-africa-corruption-bribes.

2. Ministry of Finance, Government of India, *The BRICS Report: A Study of Brazil, Russia, India, China, and South Africa with Special Focus on Synergies and Complementarities* (New Delhi: Oxford University Press, 2012).

3. Vivien Foster and Cecilia Briceno-Garmendia, *Africa's Infrastructure: A Time for Transformation,* Africa Development Forum (Washington, DC: World Bank, 2010).

4. "Guarantee by the Government of the Republic of South Africa in Favour of Noteholders for the Obligations of Eskom Holdings Ltd.," November 2009, http://www.eskom.co.za/OurCompany/Investors/GovernmentGuarantees/Pages/Government_Guarantees.aspx.

5. The Government of the Republic of South Africa and Eskom Holdings SOC Limited, "Amended and Restated Guarantee Framework Agreement," February 2012, http://www.eskom.co.za/OurCompany/Investors/GovernmentGuarantees/Pages/Government_Guarantees.aspx.

6. "State Capture Report Puts Malusi Gigaba and Lynne Brown at the Centre of Capturing and Weakening Denel," *Daily Maverick,* 2 February 2022, https://www.dailymaverick.co.za/article/2022-02-02-state-capture-report-puts-malusi-gigaba-and-lynne-brown-at-the-centre-of-capturing-and-weakening-denel/.

7. "Transnet's Orgy of Greed: How Billions were Paid in Kickbacks and Staff Cashed In," *City Press,* 22 April 2019, https://city-press.news24.com/News/transnets-orgy-of-greed-how-billions-were-paid-in-kickbacks-and-staff-cashed-in-20190422.

8. "The Tall Tale of Prasa's R3.5bn Locomotives," *News24,* 29 July 2019. https://www.news24.com/Columnists/AdriaanBasson/the-tall-tale-of-prasas-r35bn-locomotives-20190729.

9. Crispian Olver, *A House Divided: The Feud That Took Cape Town to the Brink* (Cape Town: Jonathan Ball, 2019); Crispian Olver, *How to Steal a City: The Battle for Nelson Mandela Bay: An Inside Account* (Cape Town: Jonathan Ball, 2017); Karl von Holdt, "The Political Economy of Corruption: Elite-Formation, Factions and Violence," (working paper, Society, Work and Politics Institute, University of the Witwatersrand, 2018).

10. Alexander Beresford, "Power, Patronage, and Gatekeeper Politics in South Africa," *African Affairs* 114, no. 455 (1 April 2015): 226–48.

11. Ben Fine and Zavareh Rustomjee, *The Political Economy of South Africa: From Minerals-Energy Complex to Industrialisation* (London: C. Hurst, 1996).

12. "Mbeki Apologises for SA Power Cuts," *Independent Online,* 12 December 2007.

13. Brian Dames, "The Lessons That Eskom Have Learnt the Hard Way," *Infrastructure News,* 17 September 2013, http://www.infrastructurene.ws/2013/09/17/the-lessons-eskom-learnt-the-hard-way/.

14. "World Bank Approves Eskom Loan," *Mail and Guardian,* 9 April 2010.

15. "World Bank Approves Eskom Loan."

Notes to Pages 107–110 ∽ 149

16. "Eskom Will Get R21 Bn Loan to Build Power Station," *The Star* (Gauteng), 27 November 2009.

17. "Management Response to the Request for Verification of Compliance to Bank Policies Regarding the Medupi Power Project, South Africa," *African Development Bank,* 7, https://www.afdb.org/en/documents/document /south-africa-management-response-to-the-request-for-verification -of-compliance-to-bank-policies-regarding-the-medupi-power-project -management-response-n-rq2010-2-24333.

18. Lawrence Mushwana, *Report on an Investigation into an Allegation of Improper Conduct by the Former Chairperson of the Board of Directors of Eskom Holdings Limited, Mr. V. Moosa, Relating to the Awarding of a Contract,* Report No: 30 of 2008/9 (Pretoria: Office of the Public Protector, 2009), 8.

19. Mushwana, 35.

20. "Hitachi Fined in US for Paying ANC 'Front,'" *Business Day,* 29 September 2015.

21. Securities and Exchange Commission vs. Hitachi ltd., United States District Court for the District of Columbia, 2015, 13.

22. Senior Eskom official, interview with author, 2015, Sandton.

23. "From Hitachi Deal with Arm of ANC to Power Cuts," *Bloomberg,* 5 October 2015.

24. "Hitachi and Steinmüller at Odds over SA boiler Reference Sites," *Engineering News,* 5 December 2007.

25. Interview with author (anonymous), 25 June 2015, Johannesburg Country Club.

26. "Eskom Says Medupi Welding Faults Were 'Concealed,'" *Business Day,* 19 March 2013.

27. "Eskom Says Medupi Welding Faults Were 'Concealed.'"

28. "Skills Shortage Inhibiting Success of Local Infrastructure Projects— Hitachi," *Engineering News,* 12 September 2013.

29. "Medupi Contractor Charged with Fraud," *Business Day,* 25 July 2013.

30. "Police Investigating Fraud over Welding Work at Medupi," *Business Day,* 11 September 2013.

31. "Hitachi Confirms Medupi Progress, Insists It Can Meet Revised Schedule," *Engineering News,* 14 November 2013.

32. "Hitachi Confirms Medupi Progress."

33. "Hitachi Confirms Medupi Progress."

34. "Hitachi Confirms Medupi Progress."

35. "Eskom Tries to Keep Hitachi in Line," *Engineering News,* 11 September 2013.

36. "Eskom Holdings Soc. Limited v. Hitachi Power Africa (Pty) Ltd. & Another," *Supreme Court of Appeal,* 12 September 2013.

37. Transcripts and video recordings of the hearings of the Judicial Commission of Inquiry into Allegations of State Capture can be found here: https:// www.statecapture.org.za, last accessed April 2021.

38. Daniel Leseja Marokane Affidavit submitted to the Judicial Commission of Inquiry into Allegations of State Capture, Day 277, Exhibit U15, https://www.statecapture.org.za/site/hearings/date/2020/10/6, accessed March 2021.

39. "Minister Brown's PA and the Dubai Trip Mystery," *Times Live,* 25 July 2017, https://www.timeslive.co.za/politics/2017-07-25-exclusive-minister-browns-pa-and-the-dubai-trip-mystery/.

40. Daniel Leseja Marokane Affidavit, 3.

41. Daniel Leseja Marokane Affidavit, 4.

42. Tshediso John Matona, Annexure bundle Exhibit U13, submitted to the Judicial Commission of Inquiry into Allegations of State Capture, Day 262, U13-TJM-022; https://www.statecapture.org.za/site/hearings/date/2020/9/7, accessed March 2021.

43. Stephen Nhlapo, interview with author, October 2015, Numsa HQ Newtown, Johannesburg.

44. "Medupi Project PLA," December 2008, 92.

45. "Medupi Project PLA," 25.

46. "Medupi Project PLA."

47. "Medupi Still on Hold after Eskom Vehicle Is Torched," *The Star* (Gauteng), 18 February 2013.

48. "Strike Brings Eskom's Medupi Power Plant to a Grinding Halt," *The Star* (Gauteng), 17 January 2013.

49. "Grievances Slow Medupi Progress," *Sunday Times,* 17 February 2013.

50. "Medupi Delayed: What Went Wrong?," *Mail and Guardian,* 8 July 2013.

51. "Work Stoppages at Medupi Worry Eskom," *Business Day,* 22 February 2013.

52. "Minister Refuses Delays to Medupi," *The Star* (Gauteng), 18 March 2013.

53. "Medupi Progressing Well," *Sowetan,* 1 November 2013.

54. "Workers Set Five Cars Alight as Medupi Protest Turns Violent," *The Citizen,* 25 July 2013.

55. "Travel Fight Causes New Delay at Medupi," *Business Day,* 5 August 2013.

56. "Latest Strike Is the Last Medupi Straw," *City Press,* 11 August 2013.

57. "Inside Medupi's Labor Pains," *Fin24,* 23 August 2015, https://www.news24.com/fin24/inside-medupis-labour-pains-20150822.

CONCLUSION

1. "Ramaphosa Backs Just Transition—Insists SA's Sovereignty Not for Sale in R131bn COP26 Financing Offer," *Daily Maverick,* 25 November 2021, https://www.dailymaverick.co.za/article/2021-11-25-ramaphosa-backs-just-transition-insists-sas-sovereignty-not-for-sale-in-r131bn-cop26-financing-offer/.

2. "Mantashe Says He Is Willing to Take His Coal Fight to Court," *Business Day,* 9 November 2021, https://www.businesslive.co.za/bloomberg/news/2021-11-09-mantashe-says-he-is-willing-to-take-his-coal-fight-to-court/.

3. "Unbundling of Eskom Is on Track, Says de Ruyter," *Independent Online,* 27 May 2021, https://www.iol.co.za/business-report/companies/unbundling-of-eskom-is-on-track-says-de-ruyter-3457f124-6e62-49b6-a5d6-77c7f5c62878.

4. "Union Demands Resignation of Eskom Boss André de Ruyter," *Independent Online,* 11 December 2021, https://www.iol.co.za/sundayindependent/news/union-demands-resignation-of-eskom-boss-andre-de-ruyter-95501282-1299-4b22-a913-ddb0467c56ce.

152 ⁓ *Note to Page 129*

Bibliography

Alberts, Ben. "Presidential Address: Planning the Utilisation of South Africa's Coal Reserves." *Journal of South African Institute of Mining and Metallurgy* 87, no. 11 (November 1987): 371–95.

Ballim, Faeeza. "The Pre-History of South African 'Neo-Liberalism': The Rise and Fall of Co-Operative Farming on the Highveld." *Journal of Southern African Studies* 41, no. 6 (2 November 2015): 1239–54.

———. "The Un-Making of the Group Areas Act: Local Resistance and Commercial Power in the Small Town of Mokopane." *South African Historical Journal* 69, no. 4 (2 October 2017): 568–82.

Ballim, Y., R. Omar, and A. Ralphs. "Learning Assessments in South Africa." In *Experiential Learning around the World: Employability and the Global Economy,* edited by Norman Evans, 181–97. London: Jessica Kingsley, 2000.

Bandama, Foreman, Shadreck Chirikure, and Simon Hall. "Ores Sources, Smelters and Archaeometallurgy: Exploring Iron Age Metal Production in the Southern Waterberg, South Africa." *Journal of African Archaeology* 11, no. 2 (2013): 243–67.

Barry, Andrew. *Material Politics: Disputes along the Pipeline.* Chichester, UK: John Wiley & Sons, 2013.

Bates, Robert. *When Things Fell Apart: State Failure in Late-Century Africa.* Cambridge: Cambridge University Press, 2008.

Beinart, William. *Twentieth-Century South Africa.* Oxford: Oxford University Press, 2001.

Beinart, William, and Saul Dubow. *The Scientific Imagination in South Africa: 1700 to the Present.* Cambridge: Cambridge University Press, 2021.

Beresford, Alexander. "Power, Patronage, and Gatekeeper Politics in South Africa." *African Affairs* 114, no. 455 (1 April 2015): 226–48. https://doi.org/10.1093/afraf/adu083.

Bezuidenhout, Andries, and Sakhela Buhlungu. "From Compounded to Fragmented Labour: Mineworkers and the Demise of Compounds in South Africa." *Antipode* 43, no. 2 (1 March 2011): 237–63.

Bond, P. *Elite Transition: From Apartheid to Neoliberalism in South Africa.* London: Pluto, 2000.

Boone, Catherine. *Political Topographies of the African State: Territorial Authority and Institutional Choice.* Cambridge: Cambridge University Press, 2003.

Bourdieu, Pierre. "Utopia of Endless Exploitation: The Essence of Neoliberalism." Translated by Jeremy J. Shapiro. *Le Monde diplomatique* (English ed.), December 1998.

Bozzoli, G. R. *Forging Ahead: South Africa's Pioneering Engineers.* Johannesburg: Witwatersrand University Press, 1997.

Breckenridge, Keith. "Verwoerd's Bureau of Proof: Total Information in the Making of Apartheid." *History Workshop Journal* 59, no. 1 (20 March 2005): 83–108.

Callon, Michel, and Bruno Latour. "Unscrewing the Big Leviathan: How Actors Macro-Structure Reality and How Sociologists Help Them to Do So." In *Advances in Social Theory and Methodology: Toward an Integration of Micro- and Macro-Sociologies,* edited by Karin Knorr Cetina and Aaron Victor Cicourel, 277–303. Boston: Routledge and Kegan Paul, 1981.

Callon, Michel, Pierre Lascoumes, and Yannick Barthe. *Acting in an Uncertain World: An Essay on Technical Democracy.* Translated by Graham Burchell. Cambridge, MA: MIT Press, 2009.

Chirikure, Shadreck. *Metals in Past Societies: A Global Perspective on Indigenous African Metallurgy.* Cham, Switzerland: Springer, 2015.

Christie, Renfrew. *Electricity, Industry and Class in South Africa.* St. Antony's Series. London: Macmillan, 1984.

Clark, Nancy L. *Manufacturing Apartheid: State Corporations in South Africa.* New Haven, CT: Yale University Press, 1994.

Collier, Stephen J. *Post-Soviet Social: Neoliberalism, Social Modernity, Biopolitics.* Princeton, NJ: Princeton University Press, 2011.

Conradie, Steve R., and L. J. M. Messerschmidt. *A Symphony of Power: The Eskom Story.* Johannesburg: Chris Van Rensburg, 2000.

Cooper, Frederick. *Africa since 1940: The Past of the Present.* New Approaches to African History. Cambridge: Cambridge University Press, 2002.

Davies, Robert, and Dan O'Meara. "Total Strategy in Southern Africa: An Analysis of South African Regional Policy Since 1978." *Journal of Southern African Studies* 11, no. 2 (1 April 1985): 183–211.

De Villiers, Willem Johannes. *Report of the Commission of Inquiry into the Supply of Electricity in the Republic of South Africa.* Pretoria: Government Printer, 1984.

Dubow, Saul. *Apartheid, 1948–1994.* Oxford: Oxford University Press, 2014.

———. *A Commonwealth of Knowledge: Science, Sensibility, and White South Africa, 1820–2000.* Oxford: Oxford University Press, 2006.

Eberhard, Anton. "From State to Market and Back Again: South Africa's Power Sector Reforms." *Economic and Political Weekly* 40, no. 50 (10 December 2005): 5309–17.

———. "The Political Economy of Power Sector Reform in South Africa." Working paper WP-06, Program on Energy and Sustainable Development, Stanford University, April 2004.

———. "The Political Economy of Power Sector Reform in South Africa." In *The Political Economy of Power Sector Reform: The Experiences of Five Major Developing Countries,* ed. David G. Victor and Thomas C. Heller, 215–53. Cambridge: Cambridge University Press, 2007

Edgecombe, Ruth, and Bill Guest. "The Natal Coal Industry in the South African Economy, 1910–1985." *South African Journal of Economic History* 2, no. 2 (1987): 49–70.

Edwards, Paul N., and Gabrielle Hecht. "History and the Technopolitics of Identity: The Case of Apartheid South Africa." *Journal of Southern African Studies* 36, no. 3 (1 January 2010): 619–39.

Feinstein, Charles H. *An Economic History of South Africa: Conquest, Discrimination, and Development.* Cambridge: Cambridge University Press, 2005.

Ferguson, James. "The Uses of Neoliberalism." *Antipode* 41 (1 January 2010): 166–84.

Fine, Ben. "The Minerals-Energy Complex Is Dead: Long Live the MEC." Paper presented to Amandla Colloquium, Cape Town, 15 October 2008. https://eprints.soas.ac.uk/5617/1/MineralEnergyComplex.pdf.

———. "Privatisation and the RDP: A Critical Assessment." *Transformation* 27 (1995): 2–23.

Fine, Ben, and Zavareh Rustomjee. *The Political Economy of South Africa: From Minerals-Energy Complex to Industrialisation.* London: C. Hurst, 1996.

Forrest, Kally. *Metal That Will Not Bend: National Union of Metalworkers of South Africa 1980–1995.* Johannesburg: Witwatersrand University Press, 2011.

Foster, Vivien, and Cecilia Briceno-Garmendia. *Africa's Infrastructure: A Time for Transformation.* Africa Development Forum. Washington, DC: World Bank, 2010.

Foucault, Michel. *The Birth of Biopolitics: Lectures at the Collège de France, 1978—1979.* New York: Picador, 2010.

Freund, Bill. *Twentieth-Century South Africa: A Developmental History.* Cambridge: Cambridge University Press, 2018.

Gagliardone, Iginio. *The Politics of Technology in Africa: Communication, Development, and Nation-Building in Ethiopia.* Cambridge: Cambridge University Press, 2016.

Gelb, Stephen, ed. *South Africa's Economic Crisis.* Cape Town: David Philip, 1991.

Glaser, Daryl. "The State, Capital and Industrial Decentralisation Policy in South Africa, 1932–1985." Master's thesis, University of the Witwatersrand, 1988.

Graham, Stephen, and Simon Marvin. *Splintering Urbanism: Networked Infrastructures, Technological Mobilities and the Urban Condition.* London: Routledge, 2002.

Guest, Bill. "Commercial Coal-Mining in Natal: A Centennial Appraisal." *Natalia* 18 (December 1998): 41–58.

Gumede, William Mervin. *Thabo Mbeki and the Battle for the Soul of the ANC.* London: Zed, 2007.

Bibliography ⟿ 155

Hall, P. E. "A Preliminary Note on the Waterberg Coalfield as a Possible Source of Coking Coal." *Journal of the Chemical, Metallurgical and Mining Society of South Africa* (October 1945): 124–32.

Harvey, David. *A Brief History of Neoliberalism.* Oxford: Oxford University Press, 2007.

Hecht, Gabrielle. *Being Nuclear: Africans and the Global Uranium Trade.* Cambridge, MA: MIT Press, 2012.

———. *The Radiance of France: Nuclear Power and National Identity after World War II.* Cambridge, MA: MIT Press, 1998.

Herbst, Jeffrey. *States and Power in Africa: Comparative Lessons in Authority and Control.* Princeton, NJ: Princeton University Press, 2000.

Hirschman, Albert O. *Development Projects Observed.* Washington, DC: Brookings Institution, 1967.

Hofmeyr, Isabel. "Turning Region into Narrative: English Storytelling in the Waterberg." Paper presented at conference, The Making of Class, University of the Witwatersrand, History Workshop, 1987.

Horsfall, D. W. "A General Review of Coal Preparation in South Africa." *Journal of the South African Institute of Mining and Metallurgy* 80, no. 8 (August 1980): 257–68.

Hughes, Thomas Parke. *Networks of Power: Electrification in Western Society, 1880–1930.* Reprint edition. Baltimore: Johns Hopkins University Press, 1993.

———. "The Seamless Web: Technology, Science, Etcetera, Etcetera." *Social Studies of Science* 16, no. 2 (1986): 281–92.

———. "Technological Momentum." In *Does Technology Drive History? The Dilemma of Technological Determinism,* edited by Michael L Smith and Leo Marx, 101–13. Cambridge, MA: MIT Press, 1994.

Inggs, Jon, and Stuart Jones. "An Overview of the South African Economy in the 1980s." *South African Journal of Economic History* 9, no. 1 (1 September 1994): 1–18.

Isaacman, Allen F., and Barbara S. Isaacman. *Dams, Displacement and the Delusion of Development: Cahora Bassa and Its Legacies in Mozambique, 1965–2007.* Athens: Ohio University Press, 2013.

———. "Extending South Africa's Tentacles of Empire: The Deterritorialisation of Cahora Bassa Dam." *Journal of Southern African Studies* 41, no. 3 (June 2015): 541–60.

Jacobs, Alice. *South African Heritage: A Biography of H. J. Van Der Bijl.* Pietermaritzburg, South Africa: Shuter & Shooter, 1948.

Jasanoff, Sheila. *Designs on Nature: Science and Democracy in Europe and the United States.* Princeton, NJ: Princeton University Press, 2007.

Jones, Stuart. *Banking and Business in South Africa.* New York: St. Martin's, 1988.

Keegan, Timothy. *Rural Transformations in Industrializing South Africa: The Southern Highveld to 1914.* Johannesburg: Ravan, 1986.

Larkin, Brian. "The Politics and Poetics of Infrastructure." *Annual Review of Anthropology* 42, no. 1 (2013): 327–43.

Latour, Bruno. *The Pasteurization of France.* Cambridge, MA: Harvard University Press, 1993.

Leger, Jean. "Coal Mining: Past Profits, Current Crisis?" In *South Africa's Economic Crisis,* edited by Stephen Gelb (Cape Town: David Philip, 1991).

Mains, Daniel. *Under Construction: Technologies of Development in Urban Ethiopia.* Durham, NC: Duke University Press, 2019.

Marais, Eugene. *The Road to Waterberg and Other Essays.* Cape Town: Human & Rousseau, 1972.

Marks, Shula, and Stanley Trapido. "Lord Milner and the South African State." *History Workshop,* no. 8 (1979): 50–80.

Marquard, Andrew. "The Origins and Development of South African Energy Policy." PhD diss., University of Cape Town, 2006. https://open.uct.ac.za/handle/11427/4963.

McRae, Ian. *The Test of Leadership: 50 Years in the Electricity Supply Industry of South Africa.* Muldersdrift, South Africa: EE Publishers, 2006.

Mendelsohn, Richard. *Sammy Marks: The Uncrowned King of the Transvaal.* Cape Town: David Philip, 1991.

Miescher, Stephan. "Building the City of the Future: Visions and Experiences of Modernity in Ghana's Akosombo Township." *Journal of African History* 53, no. 3 (November 2012): 367–90.

Mitchell, Timothy. *Carbon Democracy: Political Power in the Age of Oil.* London: Verso, 2011.

———. *Rule of Experts: Egypt, Techno-Politics, Modernity.* Berkeley: University of California Press, 2002.

Moodie, T. Dunbar. "The Moral Economy of the Black Miners' Strike of 1946." *Journal of Southern African Studies* 13, no. 1 (1986): 1–35.

Olver, Crispian. *A House Divided: The Feud That Took Cape Town to the Brink.* Cape Town: Jonathan Ball, 2019.

———. *How to Steal a City: The Battle for Nelson Mandela Bay: An Inside Account.* Cape Town: Jonathan Ball, 2017.

O'Meara, Dan. *Forty Lost Years: The Apartheid State and the Politics of the National Party, 1948–1994.* Johannesburg: Ravan, 1996.

Padayachee, Vishnu. "International Financial Relations." In *South Africa's Economic Crisis,* edited by Stephen Gelb, 88–109. Cape Town: David Philip, 1991.

———. "Private International Banks, the Debt Crisis and the Apartheid State, 1982–1985." *African Affairs* 87, no. 348 (1 July 1988): 361–76.

Phadi, Mosa, and Joel Pearson. *We Are Building a City: The Struggle for Self-Sufficiency in Lephalale Local Municipality.* Johannesburg: Public Affairs Research Institute, 2018. https://pari.org.za/7221-2/.

Platzky, Laurine, and Cherryl Walker. *The Surplus People: Forced Removals in South Africa.* Johannesburg: Ravan, 1985.

Posel, Deborah. "Language, Legitimation and Control: The South African State after 1978." *Social Dynamics* 10, no. 1 (1 June 1984): 1–16.

———. *The Making of Apartheid, 1948–1961: Conflict and Compromise.* Oxford: Oxford University Press, 1991.

Reid, Donald. "Industrial Paternalism: Discourse and Practice in Nineteenth-Century French Mining and Metallurgy." *Comparative Studies in Society and History* 27, no. 4 (1985): 579–607.

Rueedi, Franziska. *The Vaal Uprising of 1984 and the Struggle for Freedom in South Africa.* Woodbridge, Suffolk: James Currey, 2021.

Scott, James C. *Seeing Like a State: How Certain Schemes to Improve the Human Condition Have Failed.* New Haven, CT: Yale University Press, 1998.

Smuts, Jan. "South Africa's Social and Economic Planning Council: General Smuts's Plans for a Better and Greater Southern Africa." *Journal of the Royal African Society* 41, no. 165 (October 1942): 231–33.

Social and Economic Planning Council (South Africa). *Public Works Programme and Policy.* Edited by Hendrik Johannes Van Eck. Social and Economic Planning Council. Report, No. 10. Cape Town: Cape Times [for Government Printer], 1946.

Sparks, Stephen. "Apartheid Modern: South Africa's Oil from Coal Project and the History of a Company Town." PhD diss., University of Michigan, 2012, https://deepblue.lib.umich.edu/handle/2027.42/91528.

Star, Susan Leigh, and James R. Griesemer. "Institutional Ecology, 'Translations' and Boundary Objects: Amateurs and Professionals in Berkeley's Museum of Vertebrate Zoology, 1907–39." *Social Studies of Science* 19, no. 3 (1989): 387–420.

Steyn, Grove. "Governance, Finance and Investment: Decision Making and Risk in the Electric Power Sector." PhD diss., University of Sussex, 2001, http://ethos.bl.uk/OrderDetails.do?uin=uk.bl.ethos.390926.

———. "Investment and Uncertainty: Historical Experience with Power Sector Investment in South Africa and Its Implications for Current Challenges." Working paper, Management Programme in Infrastructure Reform and Regulation, Graduate School of Business, University of Cape Town, 15 March 2006.

Tischler, J. *Light and Power for a Multiracial Nation: The Kariba Dam Scheme in the Central African Federation.* New York: Macmillan, 2013.

Van de Walle, Nicolas. *African Economies and the Politics of Permanent Crisis, 1979–1999.* Cambridge: Cambridge University Press, 2001.

Venugopal, Rajesh. "Neoliberalism as Concept." *Economy and Society* 44, no. 2 (3 April 2015): 165–87.

Verhoeven, Harry. *Water, Civilization and Power in Sudan: The Political Economy of Military-Islamist State-Building.* Cambridge: Cambridge University Press, 2015.

Von Holdt, Karl. "The Political Economy of Corruption: Elite-Formation, Factions and Violence." Working paper, Society, Work and Politics Institute, University of the Witwatersrand, 2018.

Von Schnitzler, Antina. *Democracy's Infrastructure: Techno-Politics and Protest after Apartheid.* Princeton, NJ: Princeton University Press, 2016.

———. "Travelling Technologies: Infrastructure, Ethical Regimes, and the Materiality of Politics in South Africa." *Cultural Anthropology* 28, no. 4 (1 November 2013): 670–93.

Woolgar, Steve, ed. *Virtual Society? Technology, Cyberbole, Reality.* Oxford: Oxford University Press, 2002.

Woolgar, Steve, and Geoff Cooper. "Do Artefacts Have Ambivalence? Moses' Bridges, Winner's Bridges and Other Urban Legends in S&TS." *Social Studies of Science* 29, no. 3 (June 1999): 433–49.

Young, Crawford. *The Rise and Decline of the Zairian State.* Madison: University of Wisconsin Press, 1985.

Zulu, Nqobile. "An Analysis of the Post 1980s Transition from Pastoral to Game Farming in South Africa: A Case Study of the Marico District." PhD diss., University of the Witwatersrand, 2015.

Index

affirmative action. *See under* racial transformation

African Development Bank (AfDB), 110–11

African homelands. *See* homelands

African National Congress (ANC): anti-apartheid activities of, 65, 79, 85–86; banning of, 26; Umkhonto we Sizwe (MK), 94; ungovernability as resistance strategy of, 12

African National Congress government, 15; focus on infrastructure and technology by, 23, 106–8; Growth, Employment, and Redistribution policy of, 12, 80; impact of intraparty conflict on, 127; promise of universal electrification by, 21; Reconstruction and Development Programme of, 80, 105; white paper (1998) on energy policy, 80. *See also* corruption

African townships, 5, 9; around Ellisras, 44–45, 51–54, 63, 72; repression of ongoing protests in, 11–13, 26, 59, 75. *See also* homelands; Marapong; Mokerong

Afrikaner nationalism, 12, 31, 46, 74

Afrikaner unity, perceived betrayal of, 44

agricultural cooperatives, 48–49

air pollution, 21, 51, 61–64, 72. *See also* climate change, Eskom's contribution to

Alberts, Ben, 34

Alstom, 112, 119

apartheid regime: defense considerations of, 18, 55, 63, 67; demise of, 1, 13; hostility toward, 18, 32, 60; industrial decentralization under, 27, 41, 76; international sanctions against, 4, 26–27, 65, 67, 76, 124; social engineering by, 5, 31, 57. *See also* National

Party (NP); racial segregation under apartheid; Verwoerd, Hendrik: vision of "grand apartheid"

Atlas, 35, 77

Australia, 33, 64, 95

authoritarian high modernism, 2, 4–5, 20, 23, 41, 45, 57, 124–25. *See also* Scott, James

authoritarian rule in Africa, 6, 15, 127

bantustans. *See* homelands

Bashir, Omar al-, 7

Basic Income Grant, 14

Begg, John, 66

Beinart, William, 8, 48

Boiler Component Manufacturers (BCM), 69

Botha, P. W., 11, 44, 55, 68, 79; declaration of state of emergency by, 59, 75, 99

Botswana, 38, 46, 51, 56, 63, 67, 94

Bourdieu, Pierre, 14

Breckenridge, Keith, 8

Broederbond, 11

Brown, Lynne, 116

Business Day, 77–78, 113

Callon, Michel, 10

Cape Midlands Association, 33

Central African Federation, 6

Central Generation Undertaking (CGU). *See* centralized national grid

centralized national grid, 60–61, 83; and building of "six-pack" power stations, 61–62, 72, 75, 86, 104; extension of, 79; tariff hikes due to, 60

Christie, Renfrew, 5

climate change, Eskom's contribution to, 15, 22, 64, 110, 123, 128–29. *See also* air pollution

161

coking coal, 20, 28, 37–38; shortage of, 34–36, 39, 124

Collier, Stephen, 14, 24

color bar, 5, 28–29, 33–34, 91–92; and job reservation, 28–29, 41, 91. *See also* mechanization: impact on color bar

Commission of Inquiry into Allegations of State Capture. *See* state capture commission

Congress of South African Trade Unions (COSATU), 15, 79, 81–82, 97, 120

Conservative Party, 12, 44, 71

Cooper, Fred, 70, 74

corruption, 16, 128; flouting of procurement rules, 108–9; Transnet, 108

corruption within Eskom, 25, 106, 108, 111, 121; and improper tender processes, 18–19, 109, 111–13, 122

Crookes, Bruce, 102

Dames, Brian, 110

DB Thermal, 114

Deats, Michael, 35–38, 40, 50–51

debt crisis of 1985, 73, 75, 77

decolonization, 90–91

Deetlefs, J. P., 71

de Klerk, F. W., 72, 79

democracy, South Africa's transition to. *See* democratization

Democracy's Infrastructure: Techno-Politics and Protest after Apartheid (von Schnitzler), 9. *See also* von Schnitzler, Antina

Democratic Alliance, 111

Democratic Republic of the Congo (DRC), 7, 108

democratization, 3–4, 15, 79, 126, 129; implications for technology and infrastructure, 16, 23–24, 86, 106, 127, 129

Department of Bantu Affairs, 37

Department of Defense, 63

Department of Health, 53, 62

Department of Labour, 93

Department of Local Areas, 51

Department of Native Affairs, 51

Department of Planning and Coordination, 27

Department of Traditional Affairs, 95

depopulation of the countryside, 27, 55–56, 63

de Ruyter, Andre, 129

de Villiers, Wim, 77–78, 80, 87

dispossession of African people, 21, 24, 43, 47, 90, 105. *See also* forced removals

District Agricultural Union (DLU), 71–72

dry-cooled power stations, 65–68, 110

Dubow, Saul, 8, 31, 55

Durban Navigation Colliery (DNC). *See under* Natal collieries

Eastern Cape province, 33–34, 101, 104

Eastern Transvaal, 55, 61–64, 66, 86

Eberhard, Anton, 79–80, 82

economic freedom, 74–75, 90

economic liberalization, 3, 15. *See also* neoliberal orthodoxy

Edwards, Paul, 24, 65

Egypt, 9

Electricity Act of 1922, 1

electricity outages. *See* load shedding

electricity shortage crisis, 24, 106, 109–10, 116, 123, 127

Electricity Supply Commission of South Africa. *See* Eskom

Electricity Workers Union of South Africa (EWU), 96, 99

Ellisras (now Lephalale), 19–20, 43–44; development of town, 45–47, 50–51; hostel accommodation in, 54, 57, 93–94, 96; informal settlements around, 48, 54, 57; Iscor's presence in, 51–53, 55, 57–58; perceived security risks of, 55, 63, 65, 92; Peri-Urban Areas Board of, 51–53. *See also* dispossession of African people; forced removals; Lephalale

Engineering Week, 65–66

Eskom: ambivalent relationships between governments and, 2–4, 10–11, 19, 23, 129; archival documents of, 17; "Capital Expenditure Programme" of, 108; contradictory internal politics of, 2–3, 19; demand forecasts of, 11, 61–62, 125; deterioration of infrastructure at, 123, 126–27; housing policy review of, 70–72; indebtedness/financial crisis of, 1, 22, 75–76, 108, 122–23, 126, 129. *See also* minerals-energy complex

Eskom Housing Policy of 1995, 104–5

Eskom's expansion plan, 20–21, 61, 72, 124–25; Capital Development Fund (CDF), 60. *See also* centralized national grid; Waterberg: Eskom's entry into

162 〜 *Index*

Ethiopia, 8; Grand Ethiopian Renaissance
Dam, 108
EVT, 68–69
exclusion, 8, 101, 105

Ferguson, James, 14
financial crisis of 2007–8, 15
Financial Mail, 62, 69
Fine, Ben, 5
forced removals, 4–7, 20, 43, 45, 53, 57–58,
125; "black spot," 48, 54, 56; in Ellis-
ras/Lephalale, 47–48, 53–54, 57–58,
125. *See also* Group Areas Act of 1952;
Group Areas Board (GAB)
Forrest, Kally, 97–98
France, 69, 128; nation-building project
in, 8–9

Gagliardone, Iginio, 8
Gauteng, 21, 24, 90
Ghana, Akosombo Dam in, 7
Gigaba, Malusi, 108, 119
gold mining industry, 2, 5, 94; demand for
steel by, 30; impact on labor relations
of, 28; link between Eskom's forecasts
and, 62; and shaping of migrant
labor regime, 89, 92, 123. *See also*
migrant labor; minerals-energy complex: impact of volatile gold price on
Gordhan, Pravin, 108, 129
Graham, Stephen, 14
Griesemer, James, 10
Grootgeluk coal mine, 20–21, 35, 37–39,
41, 51–52, 59, 66–68, 70, 91, 94–95;
erection of hostel near, 57, 93–96;
selling of coal to Eskom by, 39–40,
61, 64. *See also* labor organization:
in Waterberg; Lephalale (formerly
Ellisras): trade union organization in;
Waterberg: Eskom's entry into
Grootvlei power station, 66–68
Group Areas Act of 1952, 20, 44, 72, 86, 105
Group Areas Board (GAB), 4–5, 44–47, 57,
70, 90, 125

Ham, Alec, 66
Hans Strijdom dam, 51–52
Harvey, David, 14
Hecht, Gabrielle, 8–9, 65
Herbst, Jeffrey, 7
Hertzog, J. B. M., Pact government of, 1, 29
Hirschman, Albert, 31
Hitachi, 68, 112–15, 119

Hitachi Power Africa (HPA). *See* Hitachi
homelands, 8, 11, 13, 32, 37, 41, 44, 105;
consolidation of, 53–55; KwaZulu, 32;
as pools of cheap labor, 27, 34, 89–90.
See also Lebowa
Hughes, Thomas, 8, 14, 27
hydroelectric projects, 6–7, 64, 108; TVA
as model for, 31

Ilanga power station. *See* Matimba (formerly Ilanga) power station
Industrial and Commercial Workers
Union, 98
Industrial Development Corporation, 31
industrialization, 26; link between public
health issues and, 53–54; scientific
planning as essential to, 4, 30, 41, 124;
state corporations as foundation of,
5–6, 20, 27–29, 41, 124–25
Industrial Machinery Supplies (IMS), 69
industrial revolution, South Africa. *See*
industrialization
infrastructure, evocative quality of, 11, 31
Inga-Shaba project, 7, 108
intentionality, notion of, 2, 22
International Council on Large Electric
Systems (CIGRE), 64
International Monetary Fund, 12, 15, 74, 76
iron smelting, precolonial, 29–30
Iscor, 2, 5–6, 10–12, 17–20, 27–29; ambivalent relationships between
governments and, 3, 10–11, 23, 41, 129;
contradictory internal politics of, 3,
10; demand forecasts of, 11, 20, 33–34,
59. *See also* state corporations
Iscor's infrastructure expansion plan, 20,
124–25; financial crisis linked to,
38–39, 41, 59; Newcastle steel plant,
32, 41; Sishen–Saldanha railway line,
33–34, 38. *See also* Grootgeluk coal
mine; Waterberg: Iscor's exploitation
of coalfields in

Japan, 33–34, 39, 68, 111, 115
Johannesburg, 9, 26, 35, 84–86, 94, 97–98,
109; Nigel, 113–14
Johannesburg Stock Exchange, 80, 108
Joubo Development Corporation (JDC),
51–52

Kariba Dam, 6
Kekana, Stephen, 87, 92, 96–99, 101, 104
Kendal power station, 110

Index ⁓ 163

Keynesian model of government, 11, 15, 16, 22, 30–31; impact of stagflation on, 12, 73
Koeberg nuclear power station, 61
Kruger National Park (KNP), 37, 41
Kusile power station, 108, 110, 112–13; delays in completion of, 106, 116, 121, 126, 128
KwaZulu-Natal province, 28, 32, 104

labor legislation, 91, 93, 97–98; Wiehann Commission on, 97
labor organization, 126; in Waterberg, 17, 21, 24, 90–91, 96, 126. *See also* strike action; trade unions: African
labor relations, 28, 91, 97, 102, 117–19; formalization of, 92, 96, 99; at Grootgeluk coal mine, 95–96; at Matimba power station, 17, 92, 96, 99–102
labor shortages: due to forced removals, 45, 48, 53, 72; skilled, 77, 91, 100–101, 113
labor unrest, 35, 117–21. *See also* strike action
Land Act of 1913, 43
Larkin, Brian, 11
Latour, Bruno, 10
Lebowa, 20, 45, 48, 54–55, 57, 71; rejection of so-called independence by, 56
Lephalale (formerly Ellisras), 18–19, 29, 89, 92–94, 101, 103, 105–6; Mogol Club, 40, 118; struggle for citizenship rights in, 21, 90, 121; trade union organization in, 90–91, 96, 98–99
Le Roux, Clive, 101, 103–4; contribution to racial transformation, 86–87, 99; reform of Matimba management practices by, 81–85
Liberated Metalworkers of South Africa (Limusa), 120–21
Limpopo province, 36, 120
load shedding, 15, 22, 81, 106, 109, 116, 121, 123, 127
Lombard, Jan, 74
Loots, Willie, 48

Machel, Samora, 79
Mail & Guardian, 111
Majuba power station, 62, 86, 110
Malan, D. F., 4, 31
Mantashe, Gwede, 128–29
Marais, Eugene, 37–38
Marapong, 18, 70, 94–95, 119; protests over development of, 71–72

Marikana massacre of 2012, 120
Marokane, Dan, 115–17
Marvin, Simon, 14
Maschinenfabrik Augsburg-Nürnberg (MAN), 68
Matimba (formerly Ilanga) power station, 18, 20, 39, 42, 60, 63–64, 69, 81, 106; closing of hostel at, 103–4; demands for transformation and autochthony at, 21–22, 85, 88, 99–100, 102–3, 105; National Productivity Award for, 83; reform of management practices at, 82–85; security at, 92–93, 96; technological innovation at, 65–69, 72, 82, 84–85; West Germany involvement in, 65–66, 68, 72. *See also* labor organization: in Waterberg; Lephalale (formerly Ellisras): trade union organization in; Le Roux, Clive; racial transformation: affirmative action toward
Matona, Tshediso, 116–17
Mbeki, Thabo, 80–81, 107, 109–10
McRae, Ian, 85–86
mechanization, 34–36; impact on color bar, 29, 91–92; and longwall mining, 35–36; and open-cast mining methods, 35
Medupi power station, 3–4, 14–15, 82, 94, 109–10; construction delays at, 17, 22, 24–25, 106–7, 116–17, 119–21, 126–28; cost overruns at, 108, 116, 127; erosion of Eskom's autonomy at, 17, 19, 128; irregularities in boiler work for, 113–16; Project Labour Agreement (PLA) at, 118–19; shifting alliances and imaginaries at, 16, 22; state capture operatives at, 19, 22, 107, 115, 117. *See also* dry-cooled power stations
Metal and Allied Workers Union (MAWU), 6, 96–99
Meyer, D. F., 30
Meyer, Joe, 40, 52
migrant labor, 34, 48, 71–72, 91, 94, 104; and rights of belonging, 89–90
minerals-energy complex, 5, 7, 23; collapse of, 109; impact of volatile gold price on, 13, 21, 59, 62, 70
Mitchell, Timothy, 8–9, 90
Mobutu Sese Seko, 7
modernization, 5–7, 24, 41
Mokerong, 71

Molefe, Tsholofelo, 116
Montana, Lucky, 108
Mont Pelerin Society, 13, 73
Moodie, Dunbar, 92
Moosa, Valli, 111
Mothoni, Melton, 96
Mozambique, 55, 61, 79, 91; Cahora Bassa
    Dam, 6, 61; Frelimo, 61, 79; Moatize
    coal mine, 36
Mpumalanga (formerly Eastern Trans-
    vaal), 21, 61, 106, 121
MWU-Solidarity, 99–100

Natal collieries: Durban Navigation Col-
    liery (DNC), 28, 34, 36; labor unrest
    in, 34–35
National Archives of South Africa, 17–18
National Party (NP), 11–12, 16, 30–31, 59, 75;
    factional split between *verkramptes*
    and *verligtes* in, 44, 74; prioritization
    of poor Whites by, 48–49, 75
National Union of Metalworkers of South
    Africa (Numsa), 21, 87, 97, 99–104,
    118–21, 126
National Union of Mineworkers (NUM),
    21, 96, 98, 100–104, 118, 120, 129
Ndebele, Hendrick, 95–96
neoliberal orthodoxy, 12, 21, 73–74, 77, 87,
    105; and concept of free market, 12,
    74, 87, 90, 103, 105, 126; fiscal austerity
    linked to, 3–4, 14, 23, 59; paradoxical
    role of, 13–14
neoliberal reform in South Africa, 11–12,
    21, 42, 87, 105, 126, 127; contested,
    14–15, 24, 73, 126; fiscal austerity of, 4,
    23, 76, 108–9, 121; at Matimba power
    station, 81–84, 86–88, 105. *See also*
    privatization; structural adjustment
    policies (SAPs)
Nigeria, Mambilla hydroelectric power
    station in, 108
Nixon, Richard, 33, 60, 62
Nkomati Accord of 1984, 79
Nkrumah, Kwame, 7
Northern Transvaal (now Limpopo prov-
    ince), 36, 38, 43–44, 48–50, 56–57, 120;
    commissioner of Bantu Affairs in,
    53–54; Soutpansberg, 37
nuclear power: opposition to, 64–65, 72;
    in South Africa, 32, 61
Nylstroom, 50, 57, 99

O'Flaherty, Paul, 113, 115–16, 119

oil crisis of 1973, 60; impact on state-led
    developmental project, 3, 12, 20, 24,
    32, 34, 41, 75; link between exploita-
    tion of Waterberg coalfields and, 28,
    38–39, 41–42, 124
Onverwacht, 52, 71, 95
Orange Free State, 48, 62

Padayachee, Vishnu, 75
Pan Africanist Congress (PAC), 26
paternalism, 21, 73–74, 77, 90, 92, 95; and
    transition to individual autonomy,
    85, 87, 90–92, 126
Pistorius, Johan, 49–50
poor Whites, 6, 41, 75, 124. *See also* White
    working class
Posel, Deborah, 12, 44
power, technological facet of, 7, 9, 23, 31,
    46
privatization, 11–12, 24, 76; British experi-
    ence of, 78; of Iscor, 18, 80; unbun-
    dling as process of, 14, 80
privatization and Eskom, 1, 3, 13, 21–22,
    76–77, 87, 126; Eskom Conversion
    Bill, 81; and potential problems of
    monopoly power, 78, 80–81, 124;
    unbundling plan, 82, 129

racial integration, 72, 86–87, 92
racial segregation under apartheid, 2, 4, 8,
    12, 19–21, 44, 57–58, 124–25; relax-
    ation of, 70, 74, 86. *See also* color bar
racial transformation, 4, 87–88, 102, 105;
    affirmative action toward, 99–103.
    See also Matimba (formerly Ilanga)
    power station: demands for transfor-
    mation and autochthony at
Ramaphosa, Cyril, 1, 107, 109, 127, 129
Ramodike, Magoboya Nelson, 56
Rand Daily Mail, 29, 92
Recognition of Prior Learning (RPL), 101–2
renewable energy, transition to, 110, 129
Riekert Commission of Inquiry of 1979,
    70; and recognition of urban African
    workforce, 90, 92–93
Robinson, Freek, 87
*Rule of Experts: Egypt, Techno-Politics,
    Modernity* (Mitchell), 8. *See also*
    Mitchell, Timothy
Rustomjee, Zavareh, 5

Saldanha Bay, Western Cape, 33–34, 41
Sasol, 114, 129

Index ~ 165

Scientific Imagination in South Africa, The (Beinart and Dubow), 8
Scott, James, 2, 4–5, 23, 41, 57, 124
Second World War, 8, 16, 22, 30–31, 40, 70, 73
Sekati, Sam, 47, 57
Seodisa, Lazarus, 94–95
service delivery protests, 127; prepaid meters as rallying object for, 9, 14
Sharpeville Massacre, 4, 26–27
Sieva consortium, 69, 82; litigation between Eskom and, 83–84
Sishen iron ore mine, Northern Cape, 28, 33
Sishen–Saldanha railway line. *See under* Iscor's infrastructure expansion plan
Smuts, Jan, 1, 4, 6, 28–31
Social and Economic Planning Council, 30–31
South African Communist Party, 15
South African economy, 14, 107; under apartheid, 5, 11, 31, 59, 69, 77, 125; and decline of GDP growth, 108–9, 111; junk credit status of, 109, 117
South African Railways and Harbours, 5, 29, 33, 38, 46, 108
Soweto, 85–86; uprising of 1976, 12, 21, 44, 59, 74–75
Star, 119
Star, Susan Leigh, 10
state capture, 18–19; involvement of Gupta family in, 107–9, 115–16, 128. *See also* corruption; Medupi power station: state capture operatives at
state capture commission, 17–19, 107–8, 115–17, 121
state corporations: ambivalent role of, 2–4, 6, 23, 123–25; commercial reform of, 13, 21; as government-led developmental project, 3, 6–7, 23–24, 27, 29, 31, 108, 124; institutional autonomy of, 2, 17, 23, 41, 124; private sector investment in, 11, 27, 31, 68, 75–76, 91; public sector investment in, 27, 68, 75–76. *See also* corruption; Eskom; Iscor
Steinmüller, 69, 113
Stockpoort, 46, 56
strike action, 97, 118–20; at Matimba/Grootgeluk, 93–94, 96, 102; at Medupi, 118–20; in Vaal triangle, 59; White working class, 6, 28. *See also* labor unrest

structural adjustment policies (SAPs), 12, 74, 76
Sudan, 8; Merowe Dam, 7

Tarbela dam, Pakistan, 40
technology, relationship between society and, 8–9, 23. *See also* mechanization
technopolitics, 8–10, 23, 85, 87, 117
Thabazimbi, 29, 55
Thabazimbi iron ore mine, Waterberg, 28, 35–36, 40
Thatcher, Margaret, 76
trade union organization. *See* labor organization
trade unions: African, 17; contested development of, 6, 24, 91, 96–97; as proponents of democratic socialism, 14–15, 105; role of in apartheid protest culture, 6. *See also* labor legislation; labor organization; labor relations; Lephalale (formerly Ellisras): trade union organization in; Metal and Allied Workers Union (MAWU); National Union of Metalworkers of South Africa (Numsa); National Union of Mineworkers (NUM)
trade unions, White, 86, 91–92; Boiler Makers Union, 96
Transnet. *See under* corruption
Treurnicht, Andries, 44, 71
Tshikondeni coal mine, 37
Tsotsi, Zola, 116
Tutuka power station, suspected sabotage of, 127

unemployment, 73, 118, 123, 128
United States' Security and Exchange Commission (SEC), 111–12
University of the Witwatersrand, 2, 36, 78, 98

Vaal River Colliery, 30
Vaal Triangle, 21, 24, 59, 90, 98, 126
Vaalwater, 46, 57
van der Bijl, Hendrik Johannes, 29–30
van de Walle, Nicolas, 12, 74
van Eck, Hendrik, 30–31, 66
Verhoeven, Harry, 7
Verwoerd, Hendrik, 8, 26–27; vision of "grand apartheid," 4, 6, 31, 44, 59, 125
von Schnitzler, Antina, 8–9, 13–14, 74
Vorster, John, 26, 40

Waterberg, 44; Eskom's entry into, 18, 20, 39–40, 42, 59–62, 69–70; game farming in, 49–50; hostile topography of, 37–38, 43, 49; Iscor's exploitation of coalfields in, 3, 18–20, 23, 34, 36, 38, 53, 124–25; large-scale technological systems in, 23–24, 42, 124; White settler population of, 20, 37–38, 41, 43, 46–47. *See also* Ellisras (now Lephalale); Grootgeluk coal mine; labor organization: in Waterberg; Matimba (formerly Ilanga) power station

Whelpton, John Oswald, 50–51

White South Africa, 3, 5, 10, 23, 27. *See also* Waterberg: White settler population of

white supremacy, 2, 28, 124

White working class, 6, 28–29, 41, 75. *See also* trade unions, White

Williams, Andile, 84–85

World Bank, 12, 74, 107, 110–11; Africa Infrastructure Diagnostic Report, 108

World War II. *See* Second World War

Zaire. *See* Democratic Republic of the Congo (DRC)

Zille, Helen, 111

Zimbabwe, 37, 55, 94

Zuma, Jacob, 15, 107–8, 115, 117–18, 127; allegations of corruption against, 109